Introduction to Botany

Alexey Shipunov

June 7, 2021

Shipunov, Alexey. *Introduction to Botany*. June 7, 2021 version. 192 pp. URL: http://ashipunov.info/shipunov/school/biol_154/

Title page: *Plantago major* image drawn by Alexey Shipunov.

This book was prepared at Minot State University (North Dakota, USA) with the help of students in Biology 154 and Biology 310 classes.

Contents

Foreword

While teaching botany for about twenty years, I came to the idea of re-structuring the "classical" course into a more logical sequence of themes which you will find in this textbook.

There were two main ideas that I attempted to embed here: one was to put as much plant-related information as possible into an evolutionary context, and the other was to explain complicated problems with simple words and metaphors. There are very few botany books which are trying to do the same.

One extremely important concept to understand: **plants are not animals!** Obviously, this phrase has many important meanings. First of all, since we humans are animals, it is much easier for us to understand animal life than plant life. Many terms that are associated with animal life (like "stomach" or "blood pressure") are generally well known, even intuitively. Learning botany as a beginner requires to speak about plants, and to speak, you have to learn botanical language. This is why you need to know a vast amount of terms, so be prepared to work hard.

This textbook started from students' lecture notes but now it contains much more information. All figures are either original or modified from those sources where a license allows it (e.g., Wikimedia).

If you like this book, and want to know more about plants (and also want to see more illustrations), please watch my YouTube channel, "Tales from Greenhouse", https://www.youtube.com/channel/UCxPchT-Zp8ADvsVR91HCRmA

Glossary

***K*-strategy** population growth when there is small number of offspring with high probability to survive

***r*-strategy** population growth when there is huge number of offspring with low probability to survive

absorption zone root: zone of root hairs

achene one-seeded indehiscent dry fruit of Compositae, cypsella

adventitious roots originate from stem

anatomy invisible, internal structure which needs tools like a scalpel and/or microscope to study

anomalous secondary growth when there are multiple, short lived layers of cambium

apical meristems RAM (see) and SAM (see)

apogamy apomixis (see) when an embryo develops from unfertilized gamete, parthenogenesis

apomixis making seeds without fertilization

apospory apomixis (see) when an embryo develops from the maternal diploid tissue

ataktostele vascular bundles dispersed

bipolar plant body both root and shoot systems present

botany the scientific study of plants and plant-like organisms

brachyblasts shortened shoots of pines, larches and some other Pinaceae conifers

bract scales sterile bracts under seed scales in conifers

buds embryonic shoots

bulb short, thick underground storage shoot with prevalence of leaf tissues

calciphytes plants adapted to over-presence of $CaCO_3$

Casparian strips part of endodermis cell walls which prevents apoplastic transport

central cell biggest cell of embryo sac, with two (or sometimes one) haploid nuclei

cladophylls leaf-like, flattened shoots

cleistogamous self-pollinated flowers which do not open

collenchyma living supportive tissue

companion cells nucleate "helpers" to anucleate sieve tube cells

complex tissues tissues with more than one type of cells

compound fruit fruit originated from the whole inflorescence: infrutescence

compound leaves leaves with two or more level of hierarchy

contractile roots roots which pull plant deeper in substrate

corm short, thick underground storage shoot with prevalence of stem tissues

cortex external layer of primary stem or root

cotyledon embryonic leaf

cross-pollination pollination between genetically different plants

cuticle plastic-like isolation layer

dehiscent fruits which open

dichotomous branching: when terminal bud always divides in two

double fertilization the process when two brother male gametes fertilize two sister female cells

elongation zone root: zone of expanding cells

embryo sac female gametophyte of flowering plants

endodermis the innermost layer of cortex

endophytic fungi fungi which grow inside plant body

endosperm$_1$ haploid nutrition tissue originated from female gametophyte

endosperm$_2$ triploid (sometimes diploid) nutrition tissue originated from second fertilization

epicotyl first internode of the stem

epidermis complex surface tissue

eustele vascular bundles in a ring

exodermis the outermost layer of cortex

fibers long and narrow sclerenchyma cells

fibrous root system no primary root visible

fiddleheads spiral tops of young fern leaves

floral units (FU) elements of generative system, fructifications

flower compact generative shoot with sterile, male and female zones, specifically in that order, **other flower terms see in the separate glossary in the text**

fronds leaves of ferns

fruit ripe floral unit (FU)

fusiform initials cambium cells which make vessel elements

general characters in leaf description, characters which are applicable only to the leaf as a whole

generative shoot system all generative shoots together

ground meristem primary meristem which makes cortex and pith

ground tissue same as parenchyma (see) but only applied for tissue

halophytes plants adapted to over-presence of NaCl

haustoria sucker roots of parasitic plants

heartwood non-functional part of wood

heliophytes plants adapted to full sun

hemiparasites photosynthetic plants, feeding partly on other plants

heterophylly situation when one plant has more than one leaf type

heterosporic with male and female spores

homoiohydric plants that save water

hydrophytes plants growing in water and frequently using water for the support

hygrophytes terrestrial or partly submerged plants adapted to the excess water

hypocotyl root/stem transitional place

idioblasts solitary cells dissimilar from surrounding cells

indehiscent fruits which do not open

indusia covers of groups of sporangia (sori)

inflorescence isolated generative shoot

integument extra cover of megasporangium

intercalary meristems which grow in two directions

internodes spaces between nodes

lateral meristem cambium, meristem appearing sideways

lateral veins smaller veins, typically branching out of the main vein (see)

leaf lateral photosynthetic organ of shoot with restricted growth

leaf primordia embryonic leaves

leaf scars marks of leaf petioles

leaf traces marks of leaf vascular bundles

lenticels "openings" in bark allowing for gas exchange

leptosporangia sporangia with 1-celled wall

main vein central, most visible vascular bundle of leaf (midrib)

marginal meristems which are located on margins

maturation zone root: oldest part of root

megaphyllous with leaves originated from joint branches

megasporangia female sporangia

megaspore female spore

megasporophylls modified leaves with attached megasporangia

meristems sites of cell division

merosity multiple of flower parts numbers

mesophyll photosynthetic parenchyma of leaf

mesophytes plants adapted to the average water

microspores male spores

microsporgangia male sporangia

monilophytes all Pteridophyta except lycophytes

monopodial branching: when terminal bud continues to grow every year

morphology visible, external structure

multiple fruit fruit originated from many pistils

mycoparasites plants feeding on soil fungi

mycorrhiza roots symbiotic with fungi

nodes place where leaves are attached

nucellus wall of megasporangium

ocrea part of the leaf which goes upwards along the stem

opposite leaf arrangement: two leaves per node

organ union of different tissues which have common function(s) and origin

orthotropic growth: vertical

ovule seed plants: megasporangium with integument

oxylophytes plants adapted to acidic substrates

palisade mesophyll mesophyll of elongated, tightly packed cells

parcellate reproduce vegetatively with easily rooted body parts

parenchyma tissue or cell type of spherical, roughly connected living cells

perforations openings

pericarp most of fruit tissue

pericycle parenchyma layer just outside of vascular tissues

periderm secondary dermal tissue

perisperm nutrition tissue originated from nucellus (see)

peristome mosses: attachment to moss sporangium, helps to distribute spores

petrophytes plants adapted to grow on rocky substrates

phellem external layer of periderm, cork

phelloderm internal layer of periderm

phellogen cork cambium, lateral meristem making periderm

phloem vascular tissue transporting sugars

phyllode leaf-like petioles

phyllotaxis leaf arrangement

pistil cupule, additional cover of ovules

pit structure connecting tracheids

pith central layer of primary stem or root

plagiotropic growth: horizontal

plants are not animals!

plants$_1$ all photosynthetic organisms

plants$_2$ kingdom Vegetabilia

pneumatophores air-catching heliotropic roots

poikilohydric plants that do not save water

pollen sac seed plants: microsporangium

pollen tube fungus-like cell which brings spermatia (see) to egg

pollination transfer of male gametophytes (pollen grains) from microsporangia (pollen sacs) to megasporangia (ovules) or cupules (pistils)

prickles modified, prickly stem surface growths

primary meristems intermediate tissues which start out of apical meristems and make primary tissues

primary root originates from embryo root

primary stem stem with primary tissues only

primary tissues tissues originated from RAM or SAM (optionally through intermediate meristems)

procambium intermediate meristem developing into cortex, pith and procambium, primary meristem which makes vascular tissues

protoderm primary meristem which produce epidermis or rhizodermis

protonema mosses: embryonic thread of cells

protostele central xylem surrounded with phloem

psammophytes plants adapted to grow on sandy substrates

quiescent center core part of root apical meristem

raceme basic monopodially branched inflorescence (Model I)

radial section: cross-section

RAM root apical meristem

ray initials cambium cells which make rays

rays stem: parenchyma cells arranged for horizontal transport

repetitive characters in leaf description, characters which are applicable to the leaf parts on each level of hierarchy

rheophytes water plants adapted to fast moving water

rhizodermis root epidermis, root hairs

rhizoid cells dead cells accumulating water apoplastically

rhizome underground horizontal shoot

ring porous wood: with large vessel elements mostly in early wood

root an axial organ of plant with geotropic growth

root cap protects root meristem

root nodules bulb-like structures which contain nitrogen-fixing bacteria

root pressure pressure force made solely by roots

SAM stem apical meristem

sapwood functional part of wood

schizocarp fruits which segregate into smaller indehiscent units

sciophytes plants adapted to shade

sclerenchyma dead supportive tissue

sclerophytes plants preventing water loss, they frequently employ sclerenchyma

secondary (lateral) roots originate from primary root (see)

secondary vascular tissues secondary phloem and secondary xylem

seed chimeric structure with mother (seed coat), daughter (embryo) and endosperm genotypes

seed scales megasporophylls (see) of conifers

seta mosses: stalk of the sporogon (see)

sheath part of leaf which surrounds the stem

shoot plant body unipolar body: no root system, shoots only

sieve tube cells living cells which transport sugar

simple fruit fruit originated mostly from one pistil

simple leaf leaf with one level of hierarchy

simple tissues tissues with uniform cells

siphonogamy fertilization with the help of pollen tube

solenostele vascular bundles in "hollow" cylinder

sori clusters of sporangia

spermatium aflagellate, non-motile sperm cell (plural: spermatia)

spines reduced, prickly leaves

spiral leaf arrangement, or alternate leaf arrangement: one leaf per node

spongy mesophyll mesophyll of round, roughly packed cells

sporogon moss sporophyte

stele configuration of vascular tissues in stem or root

stem axial organ of shoot

stipules small attachments to the leaf; typically, located near the base of petiole

stolon aboveground horizontal shoot

stomata (stoma) pores which opened and closed by guard cells

succulents plants accumulate water

surface / volume law when body size grows, body surface grows slower then body volume (and weight)

sympodial branching: when terminal bud degrades every year

synangia adnate sporangia

tangential section when plane is tangent to surface

tap root system primary root well developed

tendrils organ modifications using for climbing

terminal characters in leaf description, characters which are applicable only to the leaf terminals (leaflets)

thallus flat, non-differentiated body

thorns prickly shoots

thyrsus basic sympodially branched inflorescence (Model II)

tissue is a union of cells which have common origin, function and similar morphology

tracheary elements water-transporting dead cells

tracheids tracheary elements without perforations (openings)

transverse section: longitudinal

tuber enlarged portion of rhizome

tyloses "stoppers" for tracheary elements made by parenchyma cells, vessel element "stoppers"

vascular bundles "chords" made of xylem (inner) and phloem (outer) layers

vascular cylinder "hollow" cylinder made of xylem (inner) and phloem (outer) layers

vascular plants Pteridophyta + Spermatophyta

vascular tissues tissues which transport Shoot systemliquids

velamen absorption tissue made of dead cells

vessel members tracheary elements with preforations (openings)

wood secondary xylem, stem: everything deeper than vascular cambium

xerophytes plants adapted to the scarce water

xylem vascular tissue transporting water

Chapter 1

Introduction to the Introduction

1.1 Plants, Botany, and Kingdoms

Botany is the scientific study of plants and plant-like organisms. It helps us understand why plants are so vitally important to the world. Plants start the majority of food and energy chains, they provide us with oxygen, food and medicine.

Plants can be divided into two groups: **plants$_1$** and **plants$_2$**. Plants$_1$ contain **all photosynthetic organisms** which use light, H_2O, and CO_2 to make organic compounds and O_2. Plants$_1$ are defined *ecologically* (based on their role in nature).

Some plants$_1$ can be bacteria or even animals! One example of this a green slug, *Elysia chlorotica* (see Fig. 1.1). Green slugs collect chloroplasts from algae and use them for their entire life as food producers. Therefore, green slugs are both animals and plants$_1$.

Plants$_2$ are **all organisms from Vegetabilia kingdom**. Normally, plants$_2$ are green organisms with a stem and leaves. We can define them also as **multi-tissued, primarily terrestrial and photosynthetic eukaryotes**. This definition is *taxonomical* (based on evolution).

It is possible for the organism to be plant$_2$ but not plant$_1$ (see Fig. 1.2). Those who fall into that category, are fully parasitic plants (mycoparasites like *Pterospora*, root parasites like *Hydnora*, stem parasites like *Cuscuta*, and internal parasites like Pilostyles) which do not practice photosynthesis but have tissues, terrestrial lifestyle and originated from photosynthetic ancestors.

Figure 1.1. Green slug *Elysia chlorotica* caprures chloroplasts from the alga *Vaucheria litorea*.

Plants may be understood on several levels of organization: (from top to bottom) (a) ecosystems or taxa, (b) populations, (c) organisms, (d) organs, (e) tissues, (f) cells, (g) organelles, and (h) molecules (Fig. 1.3).

Botany is considered to be a "slice science" because it covers multiple levels of organization.

1.1.1 Taxonomy

Taxonomy, **systematics** and **classification** are terms with similar meanings; they are all about the overwhelming diversity of living organisms, for there are more than 2,000,000 species (and 300,000 of them belong to plants_2). **Phylogenetics** is a more fashionable term; it emphasizes the evolutionary history (**phylogeny**) of taxonomic groups (**taxa**).

This taxonomic organization is hierarchical. Most scientists accept seven main levels of taxonomy (**ranks**): the highest is **kingdom**, followed by **phylum**, **class**, **order**, **family**, **genus**, and lastly, **species**.

* * *

The highest rank, kingdoms are easy to understand as the *pyramid of life* (Fig. 1.4) which is divided into four levels—kingdoms. At the bottom is **Monera**, which con-

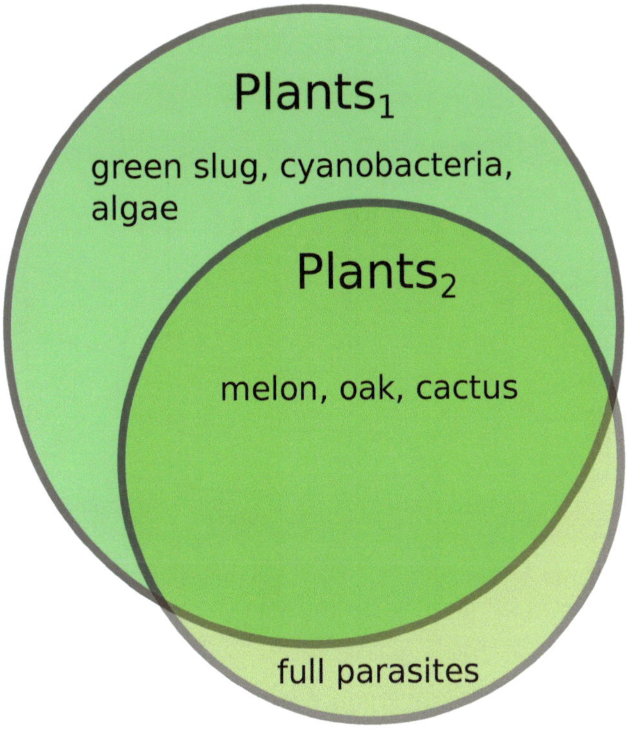

Figure 1.2. Plants$_1$ and plants$_2$.

sists of prokaryotes (Bacteria and Archaea). This is the first level of life: Monera have simplest cells without nucleus. The next level is **Protista**. These are eukaryotes (nuclear cells) without tissues; some examples are algae and fungi. The final level consists of two groups: **Vegetabilia** and **Animalia**. They both have tissues but have obtained them for completely different purposes. Animals have tissues *to hunt and digest*, while plants have tissues mainly *to survive on land*. Viri which are mentioned sideways, are not living things but merely pieces of DNA or RNA which "went astray" out of cells of living organisms of all four kingdoms. Despite of being non-living, viruses are capable of evolution.

Plants$_2$ (kingdom Vegetabilia) contain more than 300,000 species and divided in multiple subgroups (Fig. 5.1).

* * *

Ranks are used to compare taxonomic groups (**taxa**) from different major groups. No precise definitions are available for particular ranks, but it is believed that they

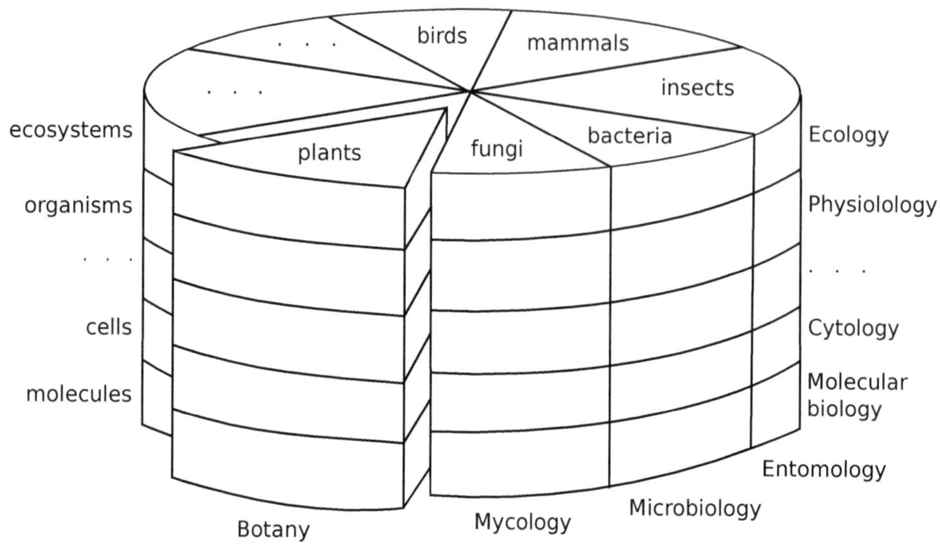

Figure 1.3. Layered pie of biology: levels of organization (left), taxonomic groups (top), "slice" sciences (bottom) and "layer" sciences (right).

are associated with the time of divergence (separation) between taxa. In addition to seven ranks mentioned above plant taxonomy uses intermediate ranks like **subfamily, subclass** or **superorder**—when taxonomic structure is too complicated.

Below is and example of names used for different ranks. Please note that names used for some ranks have standardized endings (underlined):

English	Latin	Example 1	Example 2
Kingdom	Regnum	Vegetabilia	Animalia
Phylum	Phylum	Spermatophyta	Chordata
Class	Classis	Angiospermae (Magnoliopsida)	Mammalia
Order	Ordo	Liliales	Primates
Family	Familia	Asparagaceae	Hominidae
Genus	Genus	*Chlorophytum*	*Homo*
Species	Species	*Chlorophytum comosum* (Thunb.) Jacq.	*Homo sapiens* L.

It is frequent when one species has several geographical races without clear borders between them. The example might be the stinging nettle, *Urtica dioica*. In North America, many nettles have narrower leaves and are less stinging than in Eurasia. However, there are many intermediate forms between these races. To reflect this,

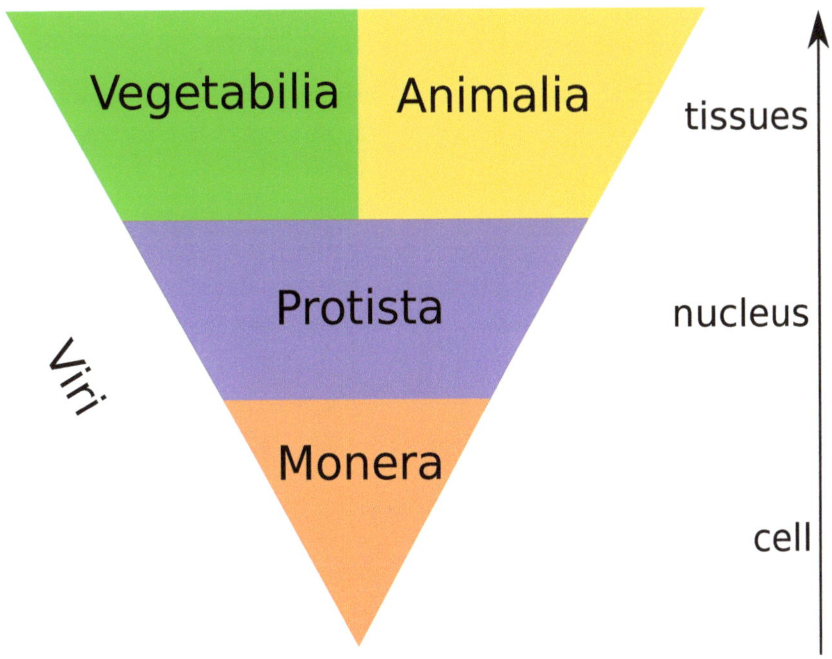

Figure 1.4. Pyramid of Life.

taxonomists introduced two **subspecies**: in this case, *Urtica diuica* subsp. *dioica* ("Eurasian") and *U. dioica* subsp. *gracilis* ("North American"). Another frequently used under-species category which is **cultivar**. Cultivars are frequently used in gardening. For example, many roses in cultivation belong to different cultivars of *Rosa banksiae*, and yellow roses are often *Rosa banksiae* cv. 'Lutea' where the last part of name is for the cultivar.

* * *

Names of species are **binomials** which consist of the name of genus and **species epithet**:

Name of species

Chlorophytum	comosum	(Thunb.)	Jacq.	1862
Name of genus	*Species epithet*	*First author*	*Second author*	*Year of description*

If one does not know the exact species, "sp." shortcut is used instead of epithet, and "spp." is used as a shortcut for multiple unknown species. It is required to use *slanted font* when one prints a name of species or genus. All scientific names are capitalized, but the second word in a species name (species epithet) always starts from lower case letter. It is a well-known fact that some species have a hybrid origin, and in these cases, botanists use a multiplication sign (×). For example, common plum (*Prunus ×domestica*) is a hybrid between blackthorn and cherry plum: *Prunus spinosa* × *Prunus cerasifera*.

The group of plants or animals must have one and only one name. Ideally, the name should be a stable ID for all occasions. But since biology is a "science of exceptions", some plant families are allowed to bear two names. As an example, legumes (**Leguminosae**) are frequently named "Fabaceae", and grasses (**Gramineae**) have the second name "Poaceae".

Throughout the long history of taxonomy, too many names were given to the same taxa. At the moment, we have almost 20,000,000 names to describe 2,000,000 species. These 18,000,000 "excess names" are **synonyms** which should not be used in science. To regulate the use of names, **nomenclature codes** were created. These codes specify, for example, the **rule of priority**: *when two names are given for the same group, only earlier name is valid.* Consequently, it is recommended to list the author and the year of description along with a name: "*Homo sapiens* L. 1758", which means that founder of taxonomy, Carolus Linnaeus ("L." shortcut) described this species in 1758.

Another important concept of nomenclature is the **nomenclature type**. Practically, this means that every species name must be associated with the physical museum specimen. In botany, these museums are collections of dried and pressed plants, called **herbaria**. Type specimens are of immense importance because there are no labels in nature, and only these specimens will "tell" about real plants or animals associated with particular names.

Names of taxa higher than species also have nomenclature types, but in these cases they are other names, not specimens. This example may clarify the use on nomenclature types. Initially, oleaster family (Elaeagnaceae) contained two genera, *Elaeagnus* (oleaster) and *Hippophaë* (sea-buckthorn). The second genus included *Hippophaë rhamnoides* (Siberian sea-buckthorn, **type species**) and *Hippophaë canadensis* (North American plant). Thomas Nuttall decided to split sea-buckthorns in two genera. Since one of them contains *Hippophaë rhamnoides*, the **type species**, it should keep the name *Hippophaë*. The second genus can be named arbitrarily. Nuttall gave it name "*Shepherdia*". As a result, the species which had name *Hippohaë canadensis* L., became *Shepherdia canadensis* (L.) Nutt.

Plant taxonomy is a science. That means that our understanding of plant groups will always change. It also means that there always are different competing opinions, the *taxonomic hypotheses* which describe plant diversity in different ways. As a result, some groups of plants could be accepted in a broad sense, including as many subgroups as possible. For example, there might be an opinion of *Homo sapiens* s.l. (*sensu lato* = wide sense) including not only contemporary humans but also Neanderthal men. As a contrast, other opinions may accept groups in a strict sense, and *Homo sapiens* s.str. (*sensu stricto* = strict sense) includes only contemporary humans.

1.2 Styles of Life and Basic Chemistry

Life obtains energy in a few different ways: (1) from sunlight (**phototrophy**); (2) from chemical reactions with inorganic matter (**lithotrophy**); (3) from breaking organic molecules into inorganic molecules, typically carbon dioxide and water (**organotrophy**). To make its body, living beings obtain building blocks either by (a) from the assimilation of carbon dioxide (**autotrophy**), or from other living beings (**heterotrophy**).

These ways combine in six lifestyles. For example, plants$_1$ are by definition **photoautotrophs**. Most plants$_2$ are also photoautotrophs, but there are exceptions: full parasites (see above). Carnivorous plants (like sundew, *Drosera* or the Venus flycatcher, *Dionaea*) are all photoautotrophs. They "eat" animals in order to obtain nitrogen and phosphorus, so the dead bodies serve not as food but as a fertilizer. Note that plants are also **organoheterotrophs** like animals because in addition to photosynthesis, all plant cells can respire.

* * *

To understand life of plants, a basic knowledge of chemistry is needed. This includes knowledge of atoms (and its components like protons, neutrons and electrons), atomic weight, isotopes, elements, the periodic table, chemical bonds (ionic, covalent, and hydrogen), valence, molecules, and molecular weight. For example, it is essential to know that protons have a charge of +1, neutrons have no charge, and electrons have a charge of −1. The atomic weight is equal to the weight of protons and neutrons. Isotopes have the same number of protons but different number of neutrons; some isotopes are unstable (radioactive).

One of the most outstanding molecules is water. Theoretically, water should boil at much lower temperature, but it boils at 100°C just because of the hydrogen bonds sealing water molecules. These bonds arise because a water molecule is *polar*: hydrogens are slightly positively charged, and oxygen is slightly negatively charged (Fig. 1.5).

Another important concept related to water is **acidity**. If in a solution of water, the molecule takes out proton (H^+), it is an **acid**. One example of this would be hy-

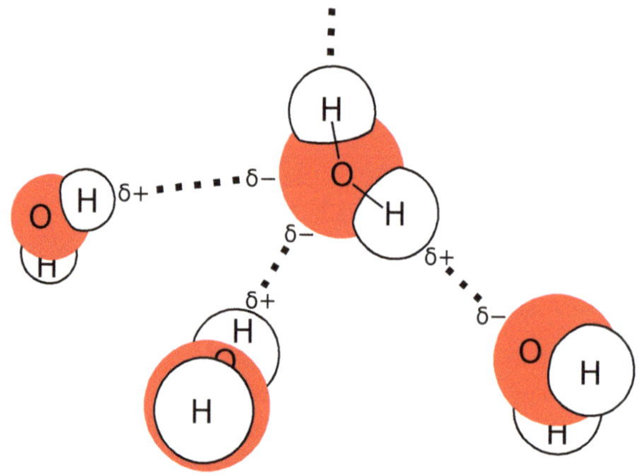

Figure 1.5. Hydrogen bonds between water molecules, δ shows the partial charge.

drochloric acid (HCl) which dissociates into H^+ and Cl^-. If the molecule takes out OH^- (hydroxide ion), this is a **base**. An example of this would be sodium hydroxide (NaOH) which dissociates into Na^+ and hydroxide ion.

To plan chemical reactions properly, we need to know about **molar mass** and **molar concentration**. Molar mass is a gram equivalent of molecular weight. This means that (for example) the molecular weight of salt (NaCl) could be estimated as 23 + 35, which equals 58 units. Consequently, one mole of salt is approximately 58 **grams**. One mole of any matter (of molecular structure) always contains $6.02214078 \times 10^{23}$ molecules (**Avogadro's number**).

The density of a dissolved substance is the **concentration**. If in 1 liter of distilled water, 58 grams of salt are diluted, we have 1M (one molar) concentration of salt. Concentration will not change if we take any amount of this liquid (spoon, drop, or half liter).

Depending on the concentration of protons in a substance, a solution can be very acidic. The acidity of a solution can be determined via pH. For example, if the concentration of protons is 0.1 M (1×10^{-1}, which 0.1 grams of protons in 1 liter of water), this is an extremely acidic solution. The pH of it is just 1 (the negative logarithm, or negative degree of ten of protons concentration). Another example is distilled water. The concentration of protons there equals 1×10^{-7} M, and therefore pH of distilled water is 7. Distilled water is much less acidic because water molecules dissociate rarely.

When two or more carbon atoms are connected, they form a **carbon skeleton**. All **organic molecules** are made of some organic skeleton. Apart from C, elements participate in organic molecules (biogenic elements) are H, O, N, P, and S. These six elements make four types of biomolecules: (1) lipids—hydrophobic organic molecules which do not easily dissolve in water; (2) carbohydrates or sugars, such as glucose (raisins contain lots of glucose) and fructose (honey); by definition, carbohydrates have multiple –OH group, there are also polymeric carbohydrates (polysaccharides) like cellulose and starch; (3) amino acids (components of proteins) which always contain N, C, O and H; and (4) nucleotides combined from carbon cycle with nitrogen (heterocycle), sugar, and phosphoric acid; polymeric nucleotides are nucleic acids such as DNA and RNA.

Chapter 2

Photosynthesis

2.1 Discovery of Photosynthesis

The history of the studies done on photosynthesis dates back into the 17th century with Jan Baptist van Helmont. He rejected the ancient idea that plants take most of their biomass from the soil. For the proof, he performed willow tree experiment. He started with a willow tree of 2.27 kg. Over 5 years, it grew to 67.7 kg. However, the weight of the soil only decreased by 57 grams. Van Helmont came to the conclusion that plants must take most of their weight from water. He did not know about gases.

Joseph Priestley ran a series of experiments in 1772 (Fig. 2.1). He tested a mouse, a candle, and a sprig of mint under hermetically sealed (no air can go in or out) jar. He first observed that a mouse and a candle behave very similarly when covered, in that they both "spend" the air. However, when a plant is placed with either the candle or mouse, the plant "revives" the air for both.

Further ideas were brought about in the late 1700's. Jan Ingenhousz and Jean Senebier found that the air is only reviving in the day time and that CO_2 is assembled by plants. Antoin-Laurent Lavoiser found that "revived air" is a separate gas, oxygen.

But what is the oxygen "maker"? There are many pigments in plants, and all accept and reflect some parts of rainbow. To identify the culprit, Thomas Engelmann ran an experiment (Fig. 2.2) using a crystal prism. He found that *Spirogyra* algae produce oxygen mostly in the blue and red parts of the spectrum. This was a huge find. It tells that the key photosynthetic pigment should accept blue and red rays, and thus reflect green rays. Blue-green **chlorophyll** best fits this description.

Another important fact was discovered by Frederick Blackman in 1905. He found that if light intensity is low, the increase of temperature actually has very little effect on

Figure 2.1. Experiments of J. Priestley (1772).

the rate of photosynthesis (Fig. 2.3). However, the reverse is not exactly true, and light is able to intensify photosynthesis even when it is cold.

This could not happen if light and temperature are absolutely independent factors. If temperature and light are components of the chain, light was first ("ignition") and temperature was second. This ultimately shows that photosynthesis has **two stages**. The first is a **light stage**. This stage relates to the intensity of the light. The second stage is the **enzymatic** (light-independent) stage which relates more with the temperature. Light reactions depend on the amount of light and water; they produce oxygen and energy in the form of ATP. Enzymatic reactions depend on carbon dioxide and water; they take energy from the light reactions and produce carbohydrates. Sometimes, enzymatic stage is called "dark" but it is not correct because in darkness, plant will run out of light-stage ATP almost immediately. Only some C_4-related processes (see below) could run at night.

* * *

Since water molecules are spent on light stage to make oxygen and at the same time are accumulating (see below), one of the best "equations" describing photosynthesis as a whole is

$$CO_2 + H_2O + \text{light} \rightarrow \text{carbohydrates} + H_2O + O_2$$

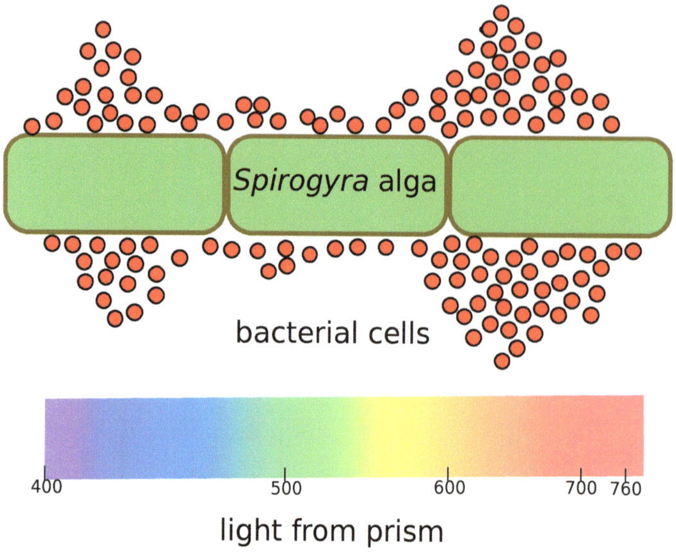

Figure 2.2. Experiment of Th. Engelmann (1881).

2.2 Light Stage

The light stage participants include photosystems ("chlorophyll"), light, water, ATPase, protons, and a hydrogen carrier ($NADP^+$).

The basic idea of light stage is that *the cell needs ATP* to assemble (later) carbon dioxide into sugar (Fig. 2.4). To make ATP, the cell needs electrical current: **proton pump**. To make this current, the cell needs the difference of electric charge (difference of potentials) between **thylakoid** (vesicle or membrane pocket) and **matrix** (stroma) compartments of the chloroplast (Fig. 2.5). To make this difference, the cell needs to segregate ions: positively charged go from outside and stay inside, negatively charged go from inside to outside. To segregate, the cell needs the energy booster—sun rays caught by the **chlorophyll** molecules embedded in the thylakoid membrane. The chlorophyll molecule is non-polar (similarly to membrane lipids) and contains magnesium (Mg). It is easy to excite the chlorophyll molecule with light; excited chlorophyll may release the electron if the energy of light is high enough.

To make carbo*hydrates* from carbon dioxide (CO_2 apparently has no hydrogen), the cell needs hydrogen atoms (H) from hydrogen carrier, NADP+ which at the end of light stage, becomes NADPH.

The main event of the light stage is that chlorophyll reacts with light, yielding electron (e^-) and becoming oxygenated, positively charged molecule. Then electron, pro-

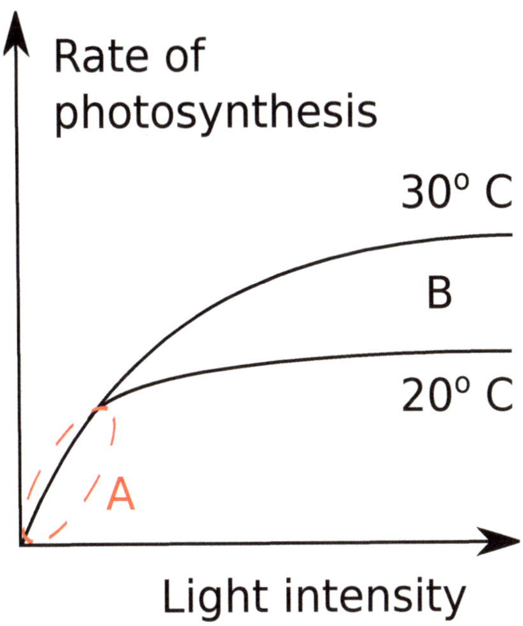

Figure 2.3. Explanation of experiment of F. Blackman (1905). A—low light, little temperature effect; B—high light, significant temperature effect.

ton and $NADP^+$ react to yield NADPH which will participate in enzymatic reactions later on. The positively charged chlorophyll is extremely active chemically, therefore it splits water molecules ("photolysis of water") into protons (which accumulate inside thylakoid), oxygen (O_2) and electron. The electron returns to chlorophyll. When increasing gradient reaches the threshold, the **proton pump** starts to work as protons (H^+) pass along the gradient. The energy of passing protons allows for the ATP synthesis from ADP and P_i (inorganic phosphate). On the other side of membrane, these protons make water with hydroxide ions.

In the previous paragraph, "chlorophyll" is actually two photosystems: photosystem II (P680) and photosystem I (P700). Photosystem II (contains chlorophyll and carotenes) is more important. It splits water, makes proton the gradient and then ATP, and forwards electrons to photosystem I. Photosystem I contains only chlorophylls and makes NADPH.

Ultimately, the light stage starts from light, water, $NADP^+$, ADP and results in an accumulation of energy (ATP) and hydrogen (NADPH) with a release of oxygen which is a kind of exhaust gas (Fig. 2.6).

Sun rays

↓

Photosystems

↓

Segregation of ions

↓

Difference of potentials

↓

Proton pump

↓

ATP

↓

CO_2 assimilation

Figure 2.4. The logical chain of light stage reactions (hydrogen carrier not shown).

2.3 Enzymatic Stage

The enzymatic stage has many participants. These include carbon dioxide, hydrogen carrier with hydrogen (NADPH), ATP, ribulose biphosphate (RuBP, or C_5), and Rubisco along with some other enzymes. Everything occurs in the matrix (stroma) of the chloroplast.

The main event of the enzymatic stage is CO_2 assimilation with C_5 into short-living C_6 molecules. Assimilation requires Rubisco as an enzyme. Next, this temporary C_6 breaks into two C_3 molecules (PGA). Then, PGA will participate in the complex set of reactions which spend NADPH and ATP as sources of hydrogen and energy, respectively; and yields (through the intermediate stage of PGAL) one molecule of glucose ($C_6H_{12}O_6$) for every six assimilated molecules of CO_2. $NADP^+$, ADP and P_i will go back to the light stage. This set of chemical reactions returns RuBP which will start the new cycle of assimilation. Consequently, all reactions described in this paragraph are part of the cycle which has the name "Calvin cycle" or "C_3 cycle" (because the C_3 PGA molecules here are most important).

In all, enzymatic stage starts with CO_2, NADPH, ATP and C_5 (RuBP). It ends with glucose ($C_6H_{12}O_6$), NADP+, ADP, P_i and the same C_5. With an addition of nitrogen and phosphorous, glucose will give all other organic molecules (Fig. 2.7).

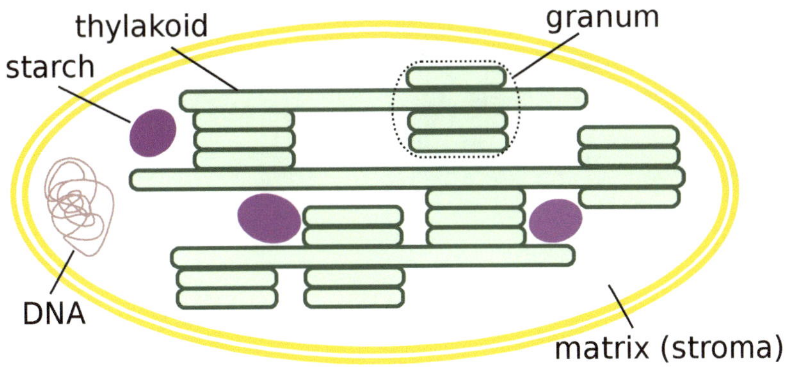

Figure 2.5. Chloroplast.

* * *

To summarize, the logic of photosynthesis (Fig. 2.8) is based on a simple idea: *make sugar from carbon dioxide*. Imagine if we have letters "s", "g", "u", and "a" and need to build the word "sugar". Obviously, we will need two things: the letter "r" and the energy to put these letters in the right order. The same story occurs in photosynthesis: it will need hydrogen (H) which is the "absent letter" from CO_2 because sugars must contain H, O and C. $NADP^+$/NADPH is used as hydrogen supplier, and energy is ATP which is created via proton pump, and the proton pump starts because light helps to concentrate protons in the reservoir.

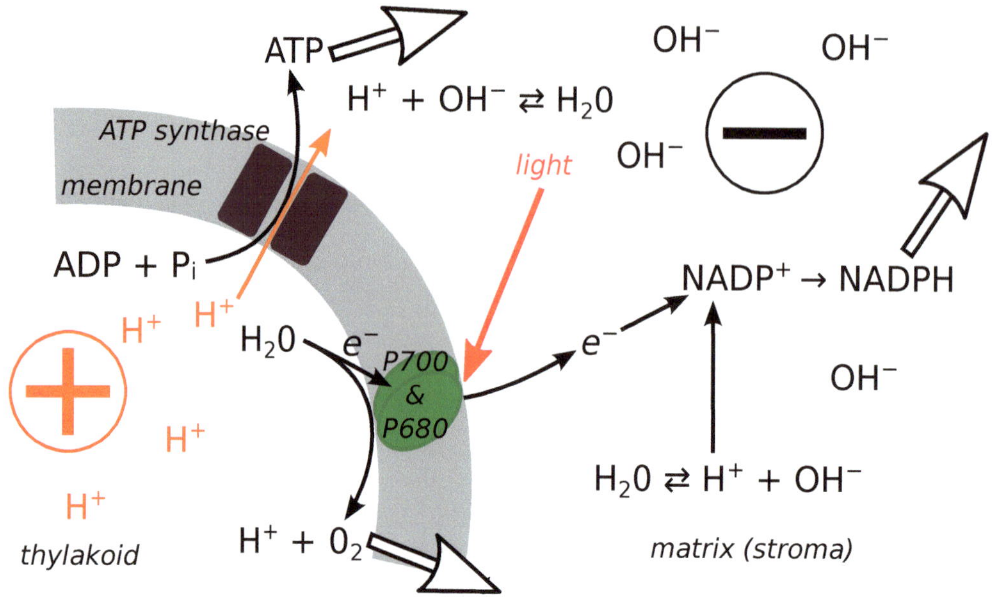

Figure 2.6. Scheme of the light stage of photosynthesis.

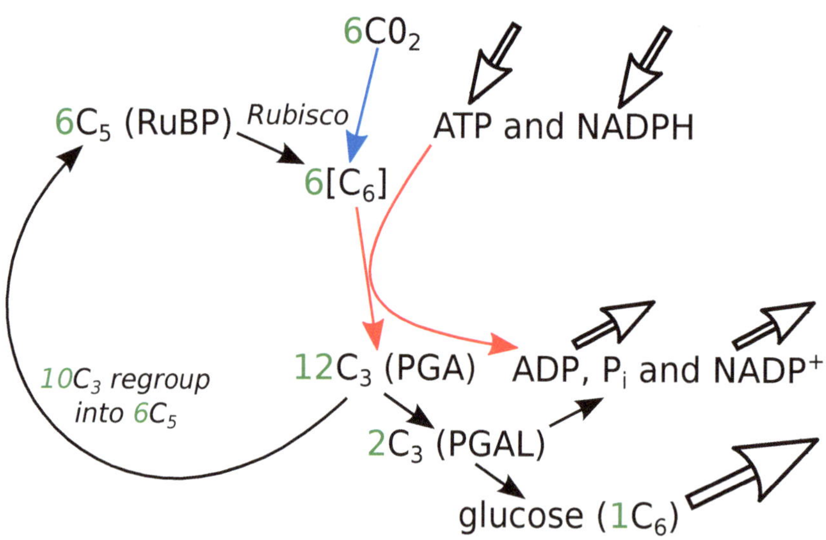

Figure 2.7. Scheme of the enzymatic stage of photosynthesis. Numbers in green show how carbon is assimilated without changing the amount of RuBP.

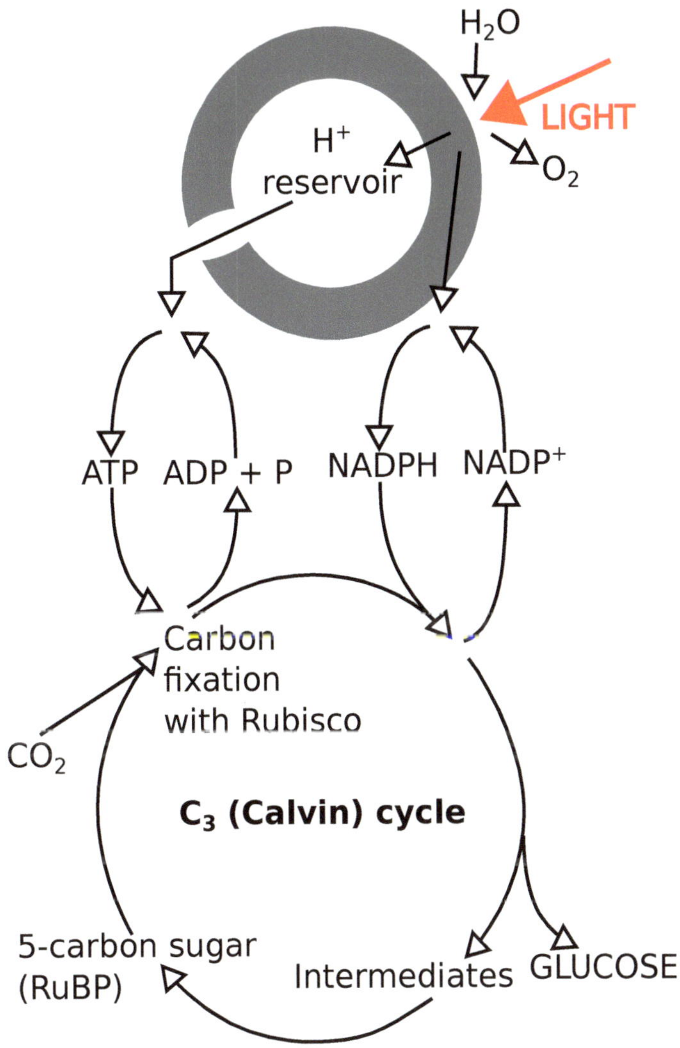

Figure 2.8. Summary of the photosynthesis. (*Based on the idea from Arms & Camp, 1986*).

2.4 C_4 Pathway

Rubisco is the enzyme of extreme importance since it starts the assimilation of carbon dioxide. Unfortunately, Rubisco is "two-faced" since it also catalyzes photorespiration (Fig. 2.9). **Photorespiration** means that plants take oxygen instead of carbon dioxide. Rubisco catalyzes photorespiration if there is a high concentration of oxygen (which usually is a result of intense light stage). Rubisco oxygenates C_5 (RuBP) which turns into PGA and PGAL, becoming glycolate. This glycolate is returned to the Calvin cycle when the cell uses peroxisomes and mitochondria, and spends ATP. The process of photorespiration wastes C_5 and ATP which could be more useful to the plant in other ways.

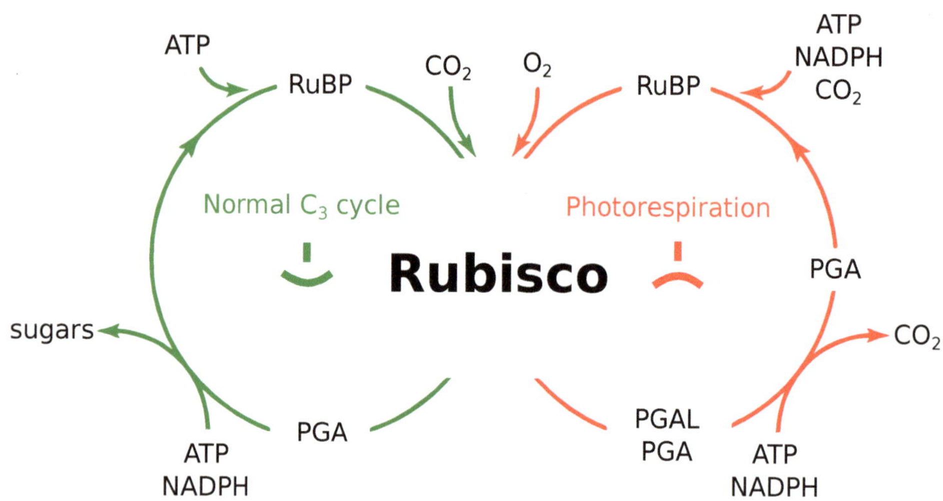

Figure 2.9. Rubisco is two-faced enzyme.

If concentration of CO_2 is high enough, assimilation will overcome photorespiration. Consequently, to minimize the amount of photorespiration and save their C_5 and ATP, plants employ **Le Chatelier's principle** ("**Equilibrium Law**") and increase concentration of carbon dioxide. They do this by temporarily bonding carbon dioxide with PEP (C_3) using carboxylase enzyme; this results in C_4 molecules, different organic acids (like malate, malic acid) with four carbons in the skeleton. When plant needs it, that C_4 splits into pyruvate (C_3) plus carbon dioxide, and the release of that carbon dioxide will increase its concentration. On the final step, pyruvate plus ATP react to restore PEP; recovery of PEP does cost ATP. This entire process is called the "**C_4 pathway**" (Fig. 2.10).

Plants that use the C_4 pathway waste ATP in their effort to recover PEP, but they still outperform photorespiring C_3-plants when there is an *intensive light and/or high*

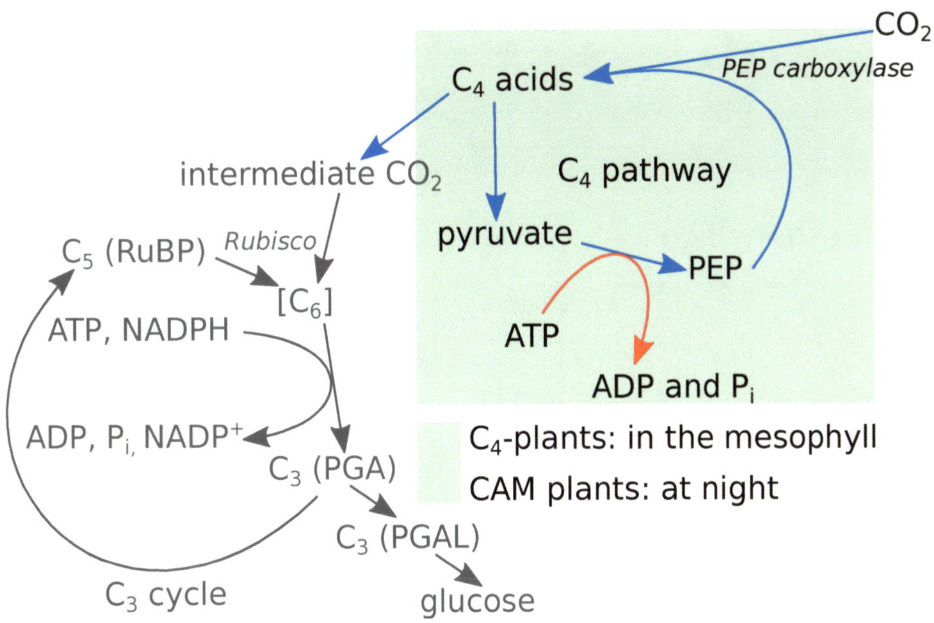

Figure 2.10. C_4 pathway (in blue).

temperature and consequently, high concentration of oxygen. This is why in the tropical climate, C_4-crops are preferable.

Two groups of plants use the C_4 pathway. Many desert or dryland plants are CAM-plants which drive the C_4 pathway at night. They make a *temporal separation* between the accumulation of carbon dioxide and photosynthesis. CAM-plants make up seven percent of plant diversity, and have 17,000 different species (for example, pineapple (*Ananas*), cacti, Cactaceae; jade plant, *Crassula* and their relatives).

"Classic" C_4 plants drive C_4 pathway in leaf mesophyll cells whereas their C_3 is located in so-called **bundle sheath cells**. This is a *spatial*, rather than temporal separation. These C_4-plants make up three percent of plant biodiversity and have more than 7,000 different species (for example, corn, *Zea*; sorghum, *Sorghum* and their relatives).

In all, both variants of C_4 pathway relate with concentration of carbon dioxide, spatial or temporal (Fig. 2.11). Both are called "carbon-concentrated mechanisms", or CCM.

There are plants which able to drive both C_3 and C_4 pathways (like authograph tree, *Clusia*), and plants having both "classic" C_4 and CAM variants (like *Portulacaria*).

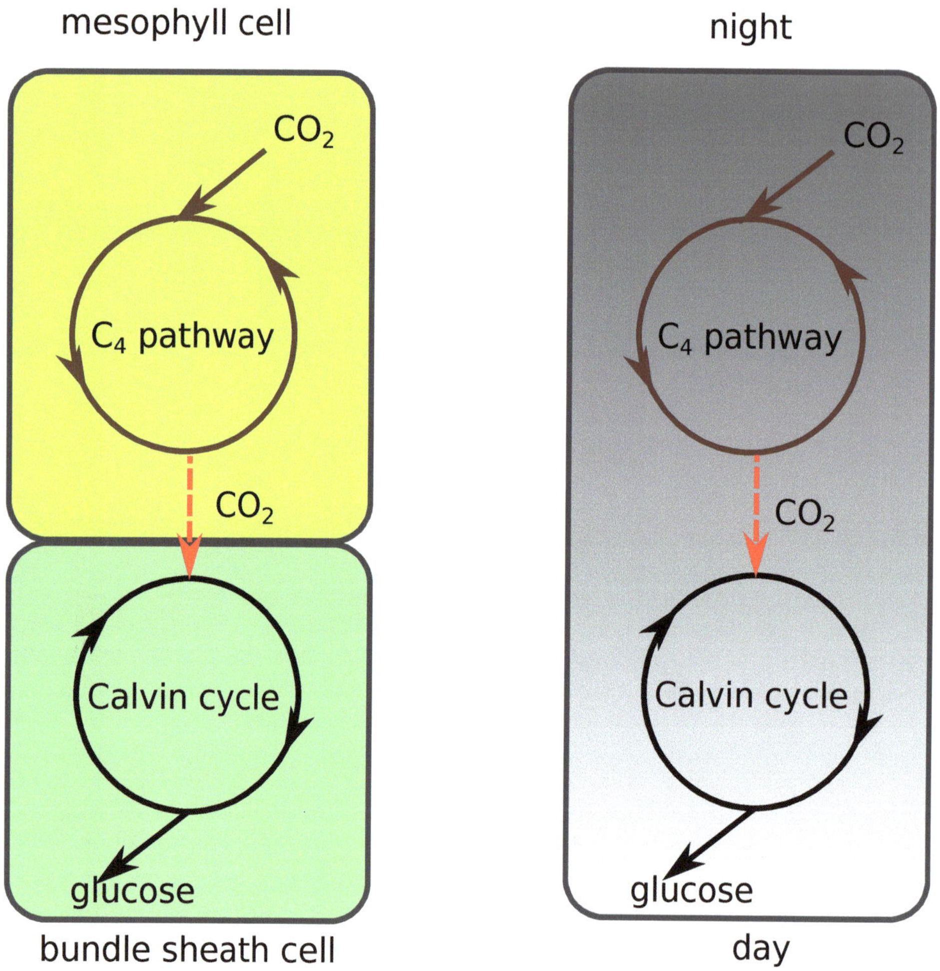

Figure 2.11. C_4 plants (left) and CAM plants (right).

2.5 True respiration

The common misconception about plants is that their only energy-related metabolic process is photosynthesis:

$$CO_2 + H_2O + \text{energy} \longrightarrow \text{carbohydrates} + O_2$$

However, as most eukaryotes, plants have mitochondria in cells and use *aerobic* (oxygen-related) respiration to obtain energy:

$$\mathbf{carbohydrates + O_2 \longrightarrow CO_2 + H_2O + energy}$$

Typically, plants spend much less oxygen in respiration than they make in photosynthesis. However, at nights plants do exactly the same as animals, and make only carbon dioxide!

Chapter 3

Symbiogenesis and the Plant Cell

3.1 Introduction to Cells

In 1665, Robert Hooke looked at cork under a microscope and saw multiple chambers which he called "cells". In 1838, Schleidern and Schwann stated that (1) *all plants and animals are composed of cells* and that (2) *cell is the most basic unit ("atom") of life*. In 1858, Virchow stated that (3) *all cells arise by reproduction from previous cells* (*"Omnis cellula e cellula"* in Latin). These three statements became the base of the **cell theory**.

Discovery of cells is tightly connected with the development of microscopy. Nowadays, there are basically three kinds of microscopy: light microscopy, transmission electron microscopy (TEM), and scanning electron microscopy (SEM). Light microscopes use normal light, it can magnify transparent things 1,000 times. Transmission electron microscopes give a more detailed view of the internal organization of cells and organelles. They use an electronic beam, which kills objects as it passes through. In addition, for examination under a TEM, objects are often stained with heavy metals like osmium, and for SEM with gold which is highly reflective for electronic rays. A TEM can magnify things 10,000,000 times. Scanning electron microscopes show an image of the surface of cells and organisms using reflected electronic beam. It can magnify things 1,000,000 times. It is possible to see atoms on these photographs!

* * *

The minimal cell should have three things: protein-synthesizing apparatus (from DNA to RNA and proteins), space designated for all other chemical reactions (jelly-like cytoplasm) and the oily film separating cell from its environment (membrane).

This is like fruit jelly covered with thin layer of butter; "fruit pieces" are protein-synthesizing parts.

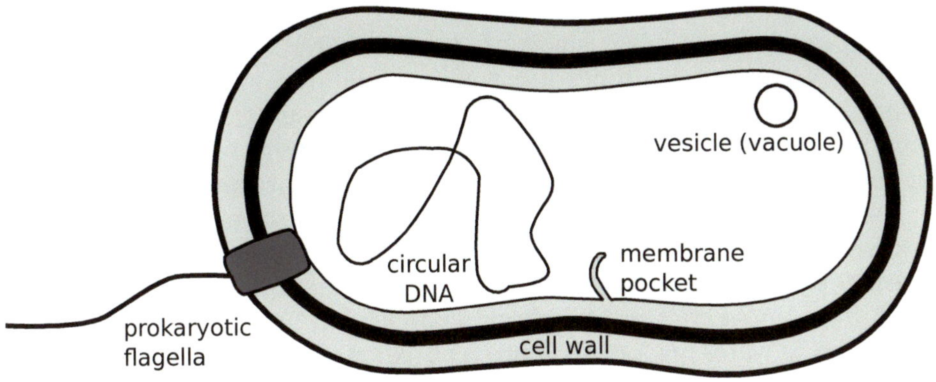

Figure 3.1. Prokaryotic cell.

The cell membrane of all cells has two layers. One end of each layer is polar and hydrophilic, while the other end is hydrophobic. These layers are made with **phospholipids** which are similar to typical lipids but have polar head with phosphoric acid, and two hydrophobic, non-polar tails (Fig. 3.2). Apart from phospholipinds, membrane contains embedded other lipids like cholesterol (in animal cells only) and chlorophyll (in some plant membranes), proteins and carbohydrates. Proteins are extremely important because without them, membrane does not allow large hydrophylic molecules and ions to came trough.

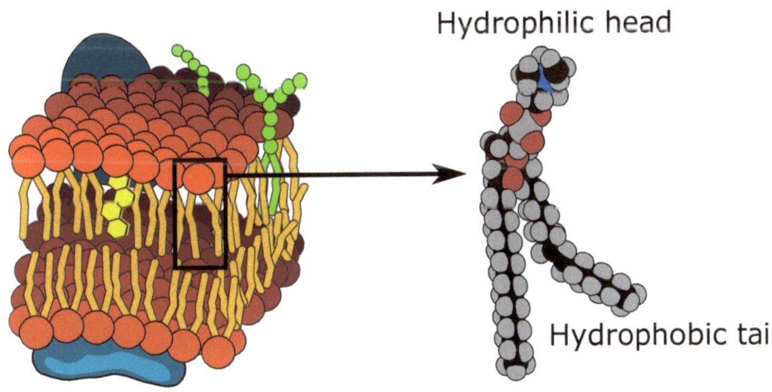

Figure 3.2. Membrane and phospholipids.

Cells which have DNA in a membrane-bound nucleus are known as **eukaryotic**, while those which do not are known as **prokaryotic**. Prokaryotic cells have their DNA surrounded by the cytoplasm. Some have also prokaryotic flagella (rotating protein structure), a cell wall, vesicles and membrane folds/pockets (Fig. 3.1). Eukaryotic cells have their DNA in a nucleus which separates it from the cytoplasm.

There are many other parts of the eukaryotic cell (Fig. 3.5). The nucleus of the cell contains DNA and proteins. Nucleoli are in the nucleoplasm, this is the place where ribosomal RNAs are assembling. Ribosomes, found in the cytoplasm, help to synthesize proteins. The **endoplasmic reticulum** (ER), usually found near edge of the cell, is where proteins are synthesized, packaged and transported. In many cells, ER is connected with nucleus membrane. The **Golgi apparatus** directs proteins and other substances to the part of the cell where they need to go. Eukaryotic cells must have **mitochondria** and might have **chloroplasts**, both originated via symbiogenesis (see below). Mitochondria are covered with two membranes, the inner membrane has intrusions called **cristae**. Mitochondria break down organic molecules into carbon dioxide and water in a process known as oxidative **respiration**.

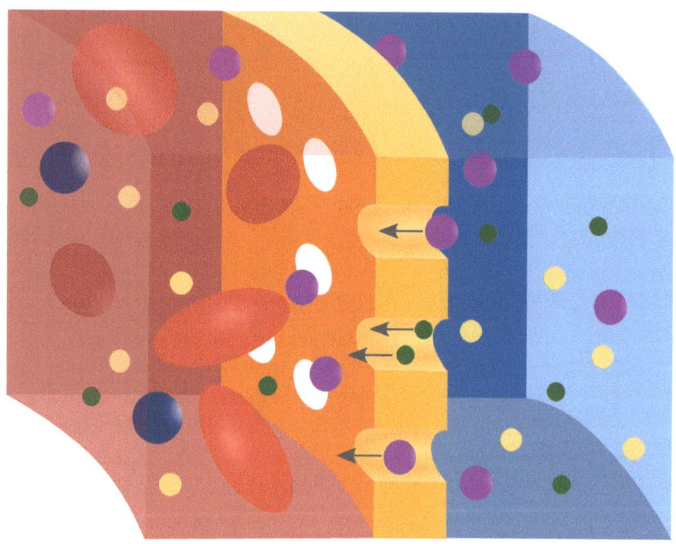

Figure 3.3. Semi-permeable membrane: how it works. Big "red" molecules are larger then pores so they are not allowed to go. Other molecules are smaller then pores and they are allowed to equalize their concentrations which are always *lower* in places where "red" molecules present. This is why they go from right to left and not otherwise.

Cell membranes are semi-permeable (Fig. 3.3), they allow some molecules (typically small and/or non-polar) to go through but others (big and/or polar) will stay outside or inside forever, or until specific pore opens. Water always "wants" to equalize concentrations on both sides of membrane and water molecules typically flow through the membrane to where concentration of other molecules (salts, acids) is *higher* (and, naturally, concentration of water is *lower*). This is **osmosis**.

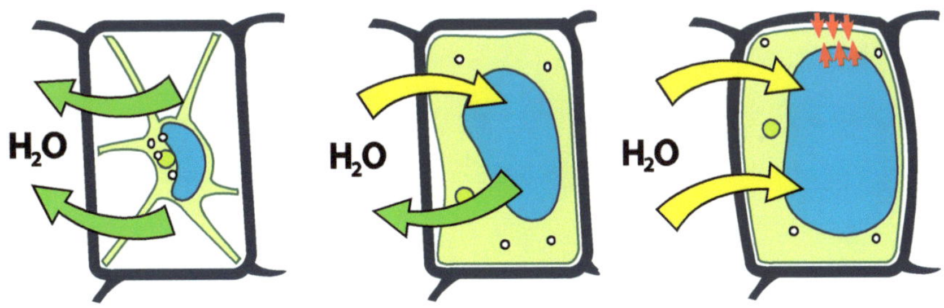

Figure 3.4. Osmosis in (from left to right) hypertonic (high salt), isotonic and hypotonic (low salt) environments. Blue color is for the vacuole. Red arrows on the right image show turgor—combined pressure of the vacuole and the cell wall.

Cell wall (common in plants and fungi) surrounds the cell and limits how far the cell can expand due to osmosis (Fig. 3.4). Since osmosis may result in uncontrollable expansion of cell, cells without cell walls must find a way to pump out the excess water. **Vacuole**(s) is the large vesicle(s) which can do a variety of things for the cell, for instance store nutrients, accumulate ions, or become a place to store wastes. It plays an important role in the *turgor* (Fig. 3.4).

3.2 Mitochondria and Chloroplasts

To escape from competition, cells which were prokaryotic became larger. To facilitate communication between all parts of this larger cell, they developed cytoplasm mobility using *actin* protein. In turn, this mobility resulted in acquiring **phagocytosis**, which is when a large cell changes shape and can engulf ("eat") other cells. This way, cells that used to be prey became predators. These predators captured prey by phagocytosis and digested bacteria in *lysosomes*, which use enzymes that destroy the cytoplasmic components of the bacterial cells.

The threat of predators result in cells became even larger, and these cells will need a better supply of ATP. Some prey which were not digested, and turned out to be useful in providing ATP. Of course, predator cells should also invent a proper transport through the resulted double membrane! Due to natural selection, those prey, which were purple bacteria, became the cell's mitochondria. This is **symbiogenesis**, or the formation of two separate organisms into a single organism (Fig. 3.6).

Another result of a larger cell (eukatyotic cells are typically 10–100 fold larger than prokaryotic) is that the size of DNA will increase, and to hold it, the cell will form a nucleus. The new predator cells also needed to prevent alien organisms from transferring their genes which will delay the evolution.

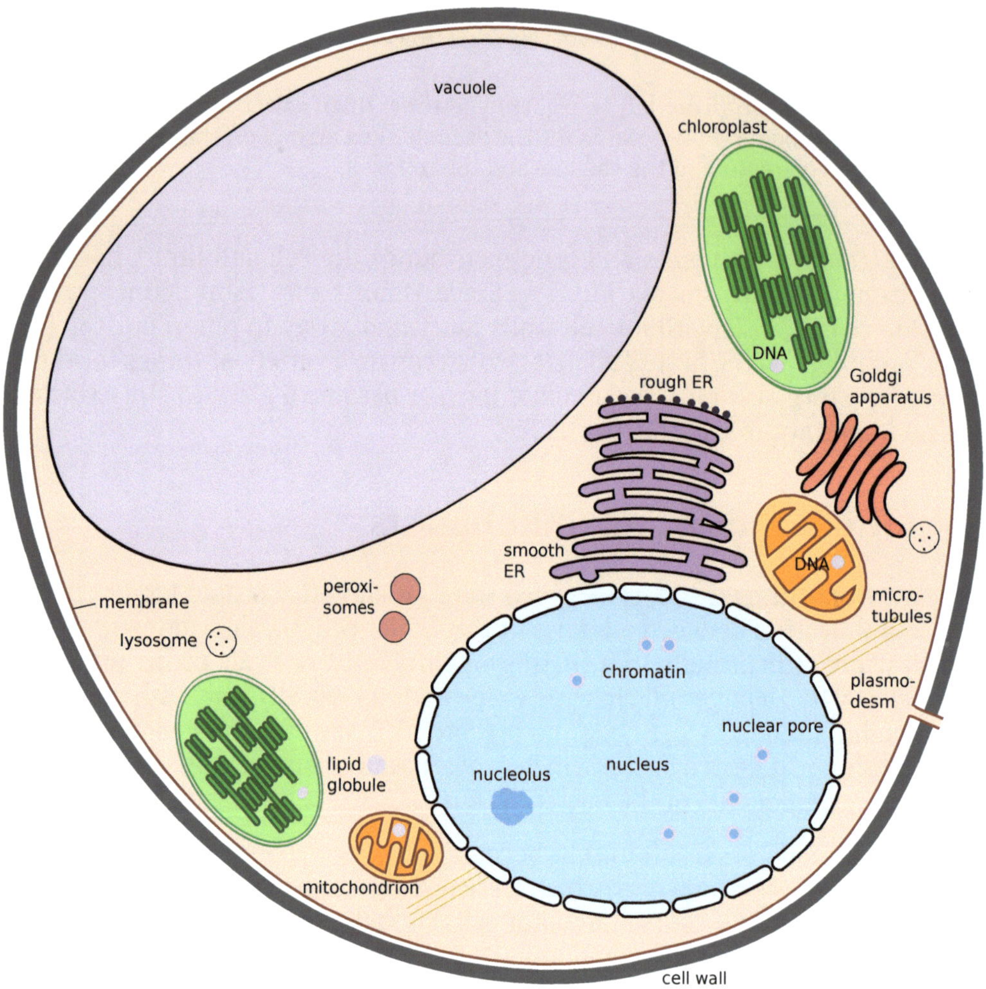

Figure 3.5. Schematic overview of the eukaryotic (plant$_1$) cell.

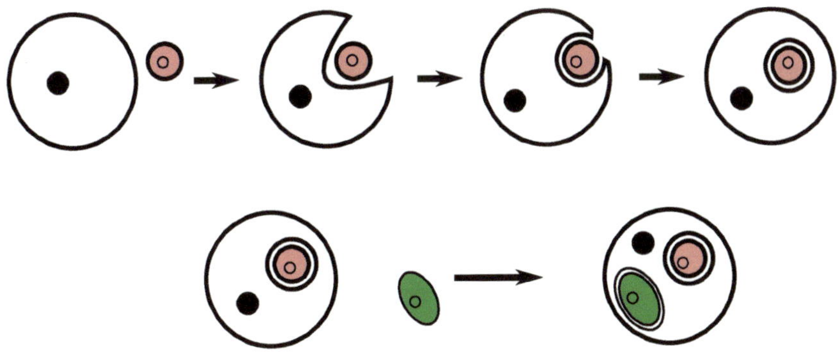

Figure 3.6. Symbiogenetic origin of eukaryotic cell (top row) and algal cell (bottom row).

The other reason is that the nucleus protects the DNA by enclosing it; in case if DNA virus comes into the cell and tries to mock up cell DNA, eukaryotic cell immediately destroys any DNA found in the cytoplasm. One more reason to make nucleus is pressure of antibiotics: nucleus improves isolation from these harmful chemicals. Nucleus formation and symbiogenesis leaded cells to become eukaryotic.

To be called an eukaryote, it is more important to have phagocytosis and mitochondria then nucleus because (1) nucleus is not always exists, it could disappear during the division of cell and (2) some prokaryotes (planctobacteria) also have membrane compartments containing DNA.

On next step, some eukaryotes also captured cyanobacteria (or another photosynthetic eukaryote), which became chloroplasts. These photosynthetic protists are called **algae**.

In all, eukaryotic cells are "second-level cells" because they are cells made up of multiple cells. Cells of all eukaryotes have two genomes, nuclear usually has biparental origin whereas mitochondial genome normally originates only from mother. Plant cells, in turn, have *three genomes*, and chloroplast genome is usually also inherited maternally.

Chloroplasts synthesize organic compounds whereas mitochondria produce most of the cytoplasmic ATP. Both organells are covered with two membranes and contain circular DNA and ribosomes similar to bacterial. Chloroplasts have thylakoids, or inner membrane pockets and vesicles. Chloroplast thylakoids could be long (**lamellae**) or short and stacked (**granes**). In turn, mitochondria could be branched and interconnecting.

Chloroplasts are normally green because of chlorophyll which converts light energy into chemical energy. Some chloroplasts lose chlorophyll and become transparent, "white", they are called **leucoplasts**. Other chloroplasts could be red and/or orange (**chromoplasts**), because they are rich of carotenes and xanthophyls. These pigments facilitate photosynthesis and are directly responsible for the fall colors of leaves. Since starch is a more compact way of storing energy than glucose, chloroplasts store carbohydrates as starch grains. Transparent **amyloplasts** contain large granules of starch. Storage tissues of potato tubers, carrot roots, sweet potato roots, and grass seeds are examples of tissues rich in amyloplasts.

Having chloroplasts and cell walls are not directly connected, but almost all organisms with chloroplasts have also cell walls. Probably, this is because cell walls do not facilitate cell motility, and for those protists which already have cell walls, obtaining chloroplast will be the nice way for coming out of competition with organotrophic beings.

3.3 Cell wall, Vacuoles, and Plasmodesmata

Among eukaryotic cells, plant cells are largest. Some of them (for example, cells from green pepper and grapefruit) are well visible with the naked eye. Plant cells do not have well-developed internal cytoskeleton, but **cell wall** provides an external one. There are two kinds (or, better, two stages of development) of cell walls, the **primary** and the **secondary**. The primary cell wall is typically flexible, frequently thin and is made of cellulose, different carbohydrates and proteins. The secondary cell wall contains also **lignin** and highly hydrophobic **suberin**. These chemicals completely block the exchange between the cell and the environment which means that the cell with secondary wall will soon die. Dead cells can still be useful to plants in many ways, for example as a defense against herbivores, support and water transport. In fact, more than 90% of wood is dead.

Since every plant cell is surrounded with a cell wall, they need a specific way of communication. This is done through **plasmodestata**—thin cytoplasmic bridges between neighbor cells. A **symplast** is the name of continuous cytoplasm inside of cells. An **apoplast** is cell walls and space outside the cell where communication and considerable metabolic activity take place. Both the symplast and apoplast are important to the transportation of nutrients needed by the cell (Fig. 3.7).

If cells are surrounded by a smaller concentration of salts than in the cytoplasm, the water will flow into the cell. This process is called **osmosis**. In plant cells, most of the water with diluted chemicals is concentrated in vacuole(s). **Turgor** pressure is the combined pressure of the cell and vacuoles wall that supports the shape of cell (Fig. 3.4).

You may think of plant tissue as about staked cardboard boxes where every box is made from wet cardboard paper (cell wall) but has the inflated balloon (vacuole)

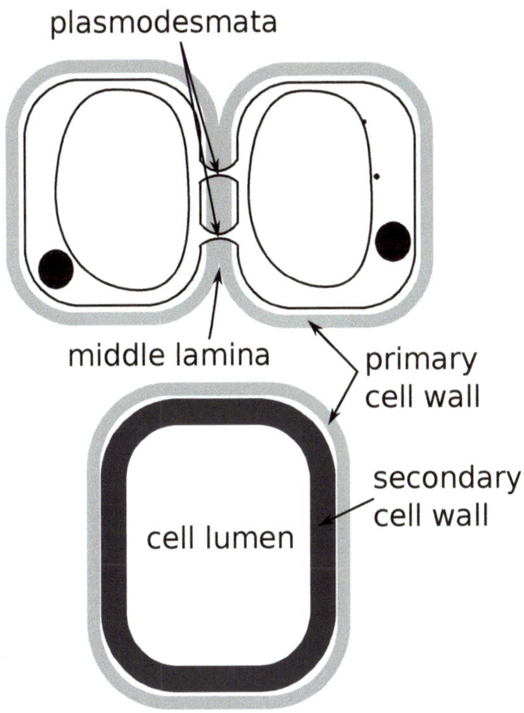

Figure 3.7. Living cells with primary cell walls (top) and dead cell with primary and secondary cell walls (bottom). Apoplast colored with shades of gray, symplast with dots.

inside, and when the pressure of vacuole decreases (water deficit), plant organs droop. Please see the video http://ashipunov.info/shipunov/school/biol_154/mov/balloon.mp4 to understand this better.

* * *

Comparing with animal cells, plant cells have chloroplasts, vacuoles, cell walls, and plasmodesmata but they hardly have any phagocytosis and true cytoskeleton (Fig 3.8). They are easy to explain: animals do not photosynthesize (no chloroplasts), instead, they need to move quickly (no cell walls and plasmodesmata); animals will support the shape of cell from cytoskeleton (no need for vacuole turgor system) and use molecular pumps to counterpart the osmosis.

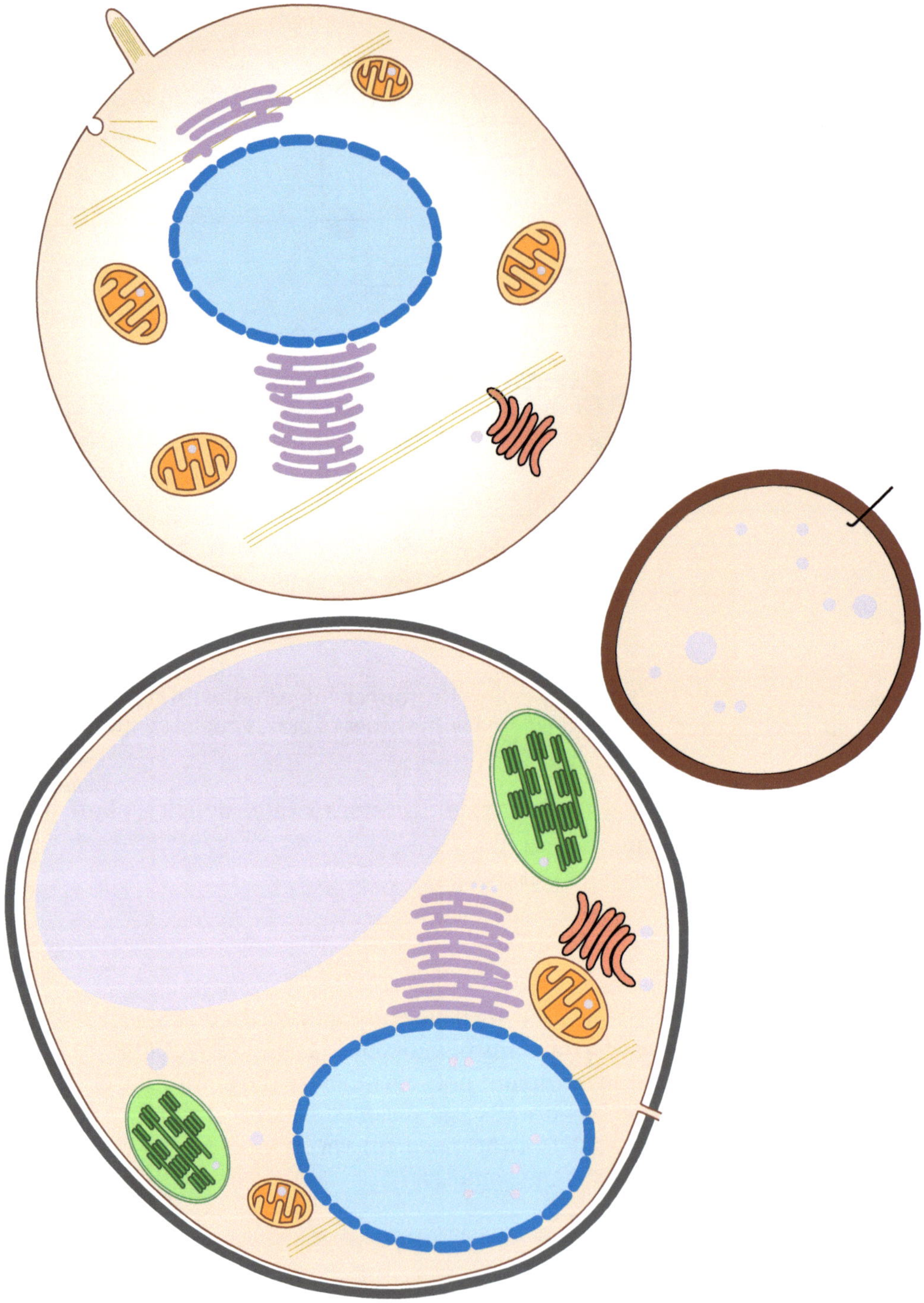

Figure 3.8. Animal, bacterial and plant cell.

3.4 Other Parts of the Cell

3.4.1 Protein Synthesis: from the Nucleus to the Ribosomes

The central dogma of molecular biology states that DNA will be converted into RNA by a process called **transcription** and RNA will be converted to protein by a process called **translation.** Translation in non-reversible whereas transcription could be reverted: there are viruses, such as HIV, that can make DNA from RNA with the enzyme called *reverse transcriptase.*

The nuclear envelope is built from a double-layered membrane. The inner and outer membranes of the nuclear envelope connect to form pores which are complicated structures controlling travel between the nucleus and the cytoplasm. Inside of the nuclear envelope there is the nucleoplasm.

Nuleoplasm contains chromatin (chromosomes). **Chromosomes** store genetic information in the form of DNA molecules. Each chromosome consists of a chain of nucleosomes, which are condensed long DNA molecules and their associated **histone** proteins. Chromatin is just another word for non-condensed chromosomes. Visible parts of chromatin (globules, filaments) correspond with non-functional DNA.

Ribosomes, which are particles that contain RNA and proteins, synthesize proteins. The **rough endoplasmic reticulum** (RER) has ribosomes along its surface, and the proteins they create are either secreted or incorporated into membranes in the cell. The **Golgi apparatus** (AG) is made of membranous sacs which are flattened and stacked, it modifies, packages, and sorts proteins and carbohydrates for the cell; this is not an essential component of cell.

3.4.2 Other Vesicles

Plant cells frequently have smaller vesicles: **lysosomes** which digest organic compounds and **peroxisomes** which, among other functions, help in photosynthesis (see above). In addition, many plant cells accumulate lipids as oil drops located directly in cytoplasm.

3.4.3 Cellular Skeleton

The cellular skeleton is a collection of protein filaments within the cytoplasm. **Microtubules** are key organelles in cell division, they form the basis for cilia and flagella and are guides for the construction of the cell wall. Cellulose fibers are parallel due to the microtubules. The movement in microtubules is based on tubulin-kinesin interactions. In contrast, the movement of **microfilaments** is based on actin-myosin interactions. Microfilaments guide the movement of organelles within the cell.

Chapter 4

Multicellularity, the Cell Cycle and the Life Cycle

4.1 Mitosis and the Cell Cycle

Mitosis is a process of equal cell division, where each of the new cells receives the same number of chromosomes as the original cell. Mitosis does not change the cells' genotype. The goal of mitosis is to distribute pre-combined genetic material equally. Actually, mitosis is a kind of **karyokinesis**, or splitting of the nucleus, as opposed to **cytokinesis**, which is the splitting of the whole cell. Karyokinesis and cytokinesis are parts of the *cell cycle* (Fig. 4.3).

All prokaryotes (Monera) have a simple cell division called "binary fission". DNA duplicates (replication), segregates and then cell splits in two (Fig. 4.1).

Eukaryotes have much more DNA than prokaryotes. This is why their cell division is more complicated. There are four stages: prophase, metaphase, anaphase, and telophase. **Prophase** is the longest, nucleus disintegrates (except some protists like fungi) and the DNA is super-spiralized into chromosomes ("archived"). In **metaphase**, the chromosomes go to the cell equator, and every "double", "X-like" chromosome is then split in two halves which schematically can be shown as

$$X \to I + I$$

In **anaphase**, microtubules move these I-like chromosomes to different poles of the cell. In **telophase**, the endoplasmic reticulum will form nuclear envelopes and DNA despiralizes (Fig. 4.2).

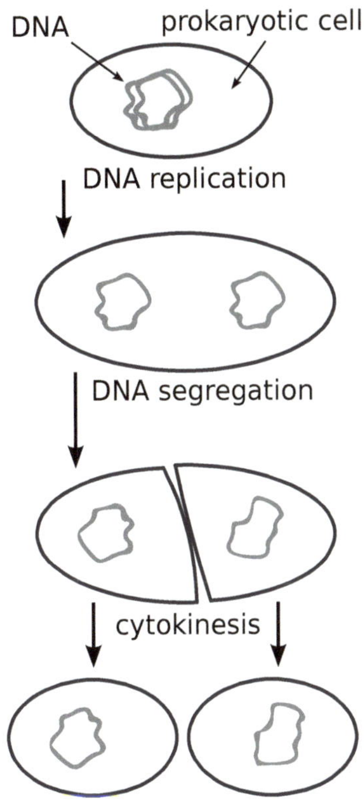

Figure 4.1. Binary fission of prokaryotes.

When mitosis is over, cell starts to divide (cytokinesis). $Plant_2$ cells use vesicles to form the border whereas many protists and animals form a constriction which finally separates two cells. Normally, chloroplasts and mitochodria are equally distributed between daughter cells along with the other cell content. Chloroplasts and mitochondria may also independently divide in "bacterial" (binary fission) way.

Mitosis is the part of the bigger **cell cycle** (Fig. 4.3). Cell cycle includes **pre-synthetic stage**, **synthetic stage**, **post-synthetic stage** (they are parts of **interphase**), karyokinesis (= mitosis) and finally cytokinesis. Apart from mitosis and cytokinesis, the most important stage of cell cycle is the **synthetic stage** (**S-stage**) when every DNA molecule (despiralized chromosome) duplicates:

$$I \rightarrow X$$

To simplify understanding of these numerous stages, one could use the following scheme. Cell cycle here has three main phases, and mitosis has four subphases:

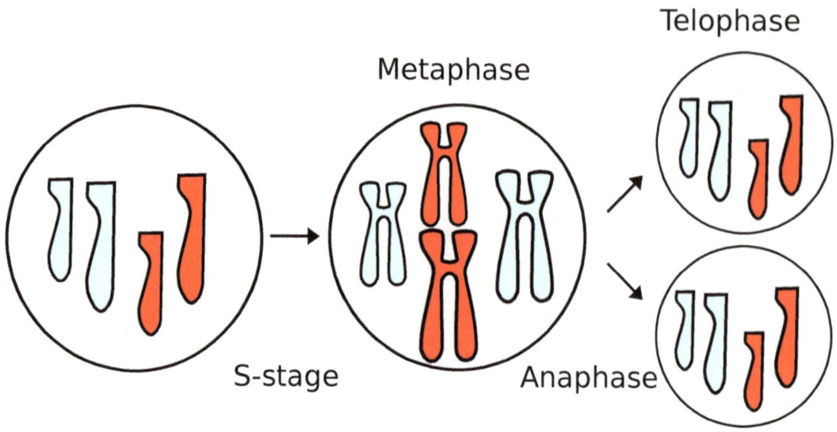

Figure 4.2. Principal scheme of mitosis.

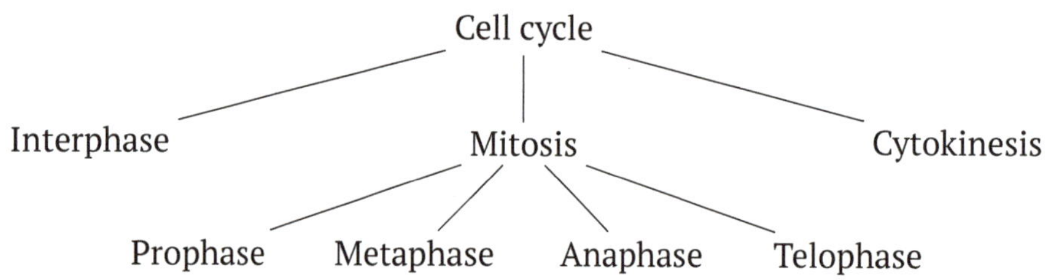

4.2 Syngamy and Meiosis

4.2.1 Sexual Process and the Syngamy

A sexual process is important to the survival of a species. First, it makes the population more diverse, which allows more flexibility to adapt via natural selection. **Natural selection** means that all organisms are different, but if environmental conditions change, only most adapted will survive. If the population is uniform, it has less chance of survival. Second, it prevents lethal mutations from being transferred to the offspring, because those with the mutations will die instead of passing on these genes. The later happens when the mutated gene is either duplicated or alone in genotype.

(**Genotype** is a gene content of the organism. A **gene** is a piece of DNA, which is equal to one protein. A **mutation** is a "mistake" in DNA. A protein (many of them are enzymes) is made up of amino acids chained together. A **population** is a group of organisms which potentially may interbreed and have no isolation barriers.)

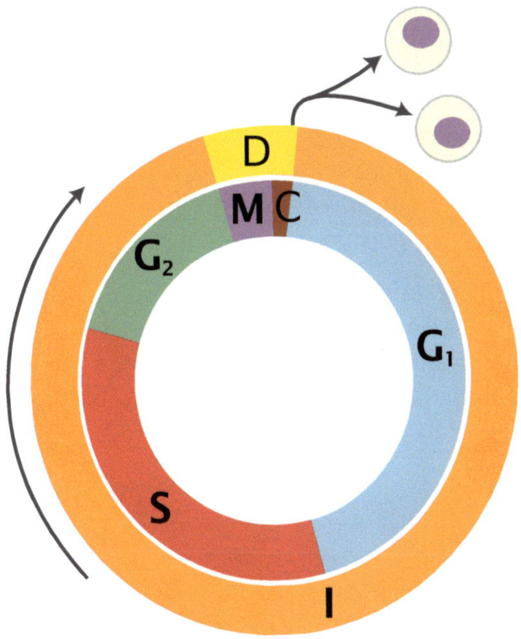

Figure 4.3. Cell cycle: I interphase, D cell division, G_1 pre-synthetic stage, S synthetic stage, G_2 post-synthetic stage, M mitosis (karyokinesis), C cytokinesis.

To make populations more diverse, organisms need to exchange DNA. One way that cells exchange genes is through syngamy. **Syngamy** (frequently labeled by "Y!"), is the fusion of two cells, resulting in a cell that has twice as many chromosomes. The two cells which are fused together are called **gametes**, and the resulting cell is a **zygote**. The goal of syngamy is the renovation of genetic material. The new cells have genotype different from the gametes. Continuous syngamy will increase the amount of DNA, so cells use meiosis (frequently labeled by "R!") to counterbalance this side-effect of syngamy:

$$\mathrm{Y!} \rightarrow \mathrm{R!}$$

Syngamy results in diploid cell:

$$X + X \rightarrow XX$$

In **diploid** organisms, chromosomes form pairs (these paired chromosomes are known as **homologous**), whereas in **halploid** organisms they remain single.

There are three types of syngamy (Fig. 4.4): isogamy, heterogamy, and oogamy. **Isogamy** happens when the gametes that fuse together are similar. To avoid self-fertilization, they must have an advanced system of recognition. Different genotypes (mating types) recognize each other with the help of surface proteins, like cells of immune system.

Heterogamy is when the gametes are of two different sizes. This difference makes recognition easier, but even more important is *division of labor*: the **female** is larger because it has resources to care for the offspring, whereas the **males** are smaller and can increase in number to allow competition and make fertilization more likely.

Oogamy is when the gametes also have different mobility. In oogamy, the non-motile female is known as the **oocyte**, and the flagellate male as the **spermatozoon**, which is only one mobile gamete here. In some organisms (red algae, sponges, crustaceans, most seed plants), spermatozoon become non-motile **spermatium** so it will need external agents to move it. Both spermatozoa and spermatia are called **sperms**.

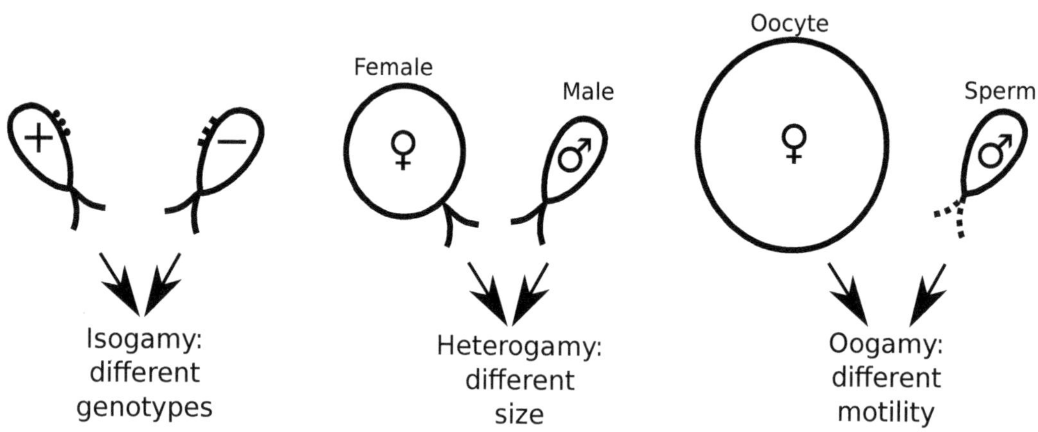

Figure 4.4. Three types of syngamy.

4.2.2 Meiosis

Syngamy is the way for organisms to become more genetically diverse, but since it increases the amount of chromosomes, it needs to be balanced by meiosis. Meiosis reduces the number of chromosomes, recombines the chromosomes, and allows chromosomes to exchange their genetic material.

Meiosis is a reductive form of cell division, where each new cell receives half of the original cell's chromosomes. Unlike mitosis, meiosis does change the genotype of cells because whole chromosomes are recombined and also exchanged their genetic material. Another difference is that in mitosis, **ploidy** ("twoness" of chromosomes) stays constant, while in meiosis, ploidy halves.

There are two problems of meiosis: first, how to find out which chromosomes are homologous; and second, how to split chromosomes which were already duplicated in S-phase. First problem is solving with "gluing" homologous chromosomes together; this happens because similar chains of DNA can attach each other. Second problem is usually solving with the second stage of meiosis which is quite similar to ordinary mitosis.

There are two stages of meiosis: a reductive division (meiosis I, unique) and an equal division (meiosis II, similar to mitosis). Each of these stages are divided into prophase, metaphase, anaphase, and telophase.

In prophase I, chromosomes *conjugate* (form **synapses**), and start to exchange DNA (crossing-over).

In anaphase I, chromosomes from each pair will *go independently to different poles*. *Independence* means that if we label "mother" and "father" chromosomes with, saying a and b, then two variants are possible:

$$X_aX_b + Y_aY_b \rightarrow (X_a + Y_a) + (X_b + Y_b)$$

or

$$X_aX_b + Y_aY_b \rightarrow (X_a + Y_b) + (X_a + Y_b)$$

because chromosomes do not know which is "father's" and which is "mother's".

Telophase I usually flows into prophase II. This second division of meiosis is very similar to mitosis *without synthetic stage* before it. Frequently, nuclei do not form until telophase II (Fig. 4.5).

In the first division, cell needs to split pairs of homologs to reduce ploidy. The second division of meiosis is necessary because DNA was already duplicated in the synthetic stage of the cell cycle. Consequently, every X-like chromosome needs to be split into two I-like chromosomes:

$$XX \rightarrow X + X \rightarrow I + I + I + I$$

This is why there are *two divisions* and *four cells* in the end (sometimes, however, only one of these four survives). If DNA would not be duplicated before, it is also possible for meiosis to happen in one stage instead of two. This kind of meiosis is described in some protists. Inverted meiosis, when reductive division is the second and equal the first, is rare but also exists in nature (e.g., in some rushes, bugs and butterflies).

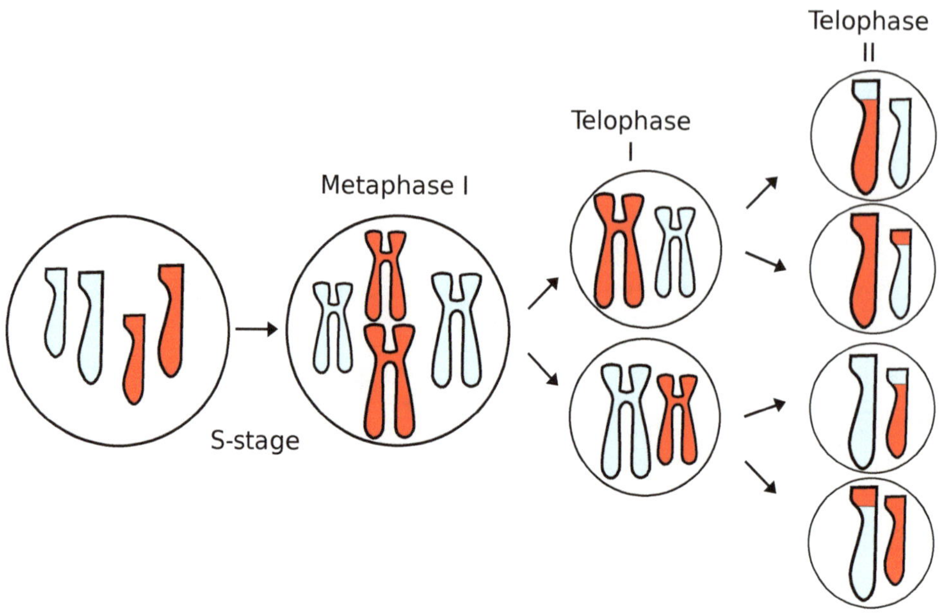

Figure 4.5. Principal scheme of meiosis. Only one of two telophase I variants is shown.

It is possible that meiosis won't work properly, which results in a cell receiving a double set of chromosomes. If, in turn, that cell goes to syngamy, the resulting zygote will have 3 sets of chromosomes. Cells with more than two sets of chromosomes are called **polyploids**. Rarely, only some chromosome pairs do not want to split. In this case, after the syngamy, some chromosomes will be "triplicated" (**trisomy**). This is **aneuploidy**. One example of frequent (1/800 births) aneuploidy in humans is Down syndrome.

4.3 Life cycle of the Unicellular Eukaryote

The life cycle of a unicellular organism begins with syngamy: one cell unites with cell having different genotype. To recognize each other, cells which are going to fuse (**gametes**) frequently use surface proteins, like cells of our immune system. If these proteins are same (same genotype), gametes will not fuse. Two fused gametes form a zygote, new diploid organism. Many unicellular protists use a zygote as a wintering stage. On spring, zygote splits with meiosis, and four haploid **spores** start four new organisms which reproduce all summer with mitosis (vegetative reproduction, cloning):

$$X + X \to XX \to I + I + I + I \to X \to I + I \to \dots$$

Despite its simplicity, this life cycle has all three possible ways of reproduction: **sexual** (ploidy doubles: syngamy), **asexual** (ploidy reduces: meiosis of zygote) and **vegetative** (ploidy does not change: mitotic divisions). To mark these ways of reproduction, we will use "**R!**" shortcut for the meiosis, and "**Y!**" shortcut for syngamy (Fig. 4.6). It should be noted that before every mitosis (and meiosis), cell DNA goes through duplication (S-stage of the cell cycle).

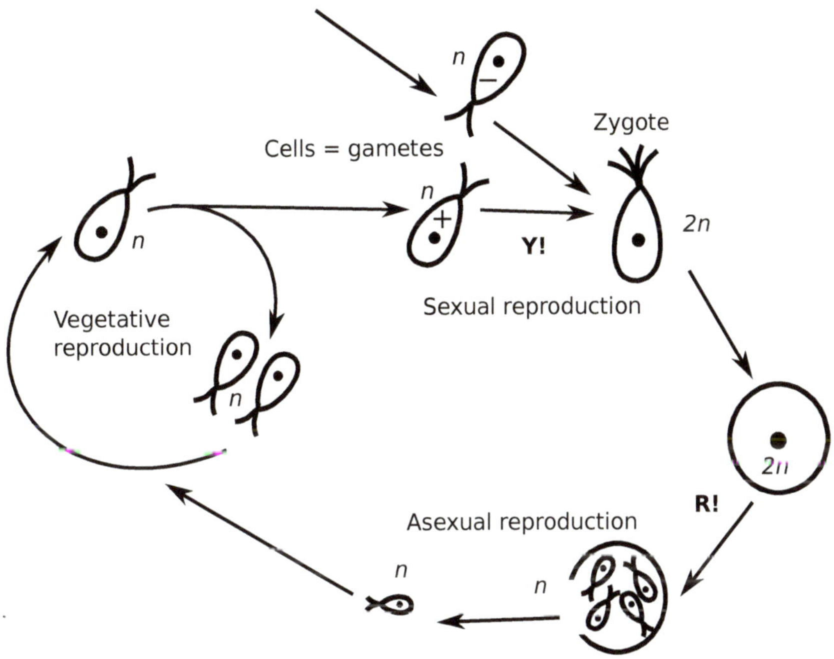

Figure 4.6. The life cycle of unicellular eukaryote.

4.4 Life cycle of the Multicellular Eukaryote

4.4.1 Origin of Death

Cells do not always part after mitosis, but sometimes stay together to form **multicellular organisms**. This increases their size, and hence provides a defense against predators. Unfortunately, it is not possible simply to increase the size of cell because the really big cell will have *less surface* (in relation to the volume), therefore it will have multiple difficulties with photosynthesis, respiration and other processes which relate with surface of cell. But many cells together will make surface big enough (Fig. 4.7). Multicellular organism has two modes of growth: scaling the body and multiplying cells.

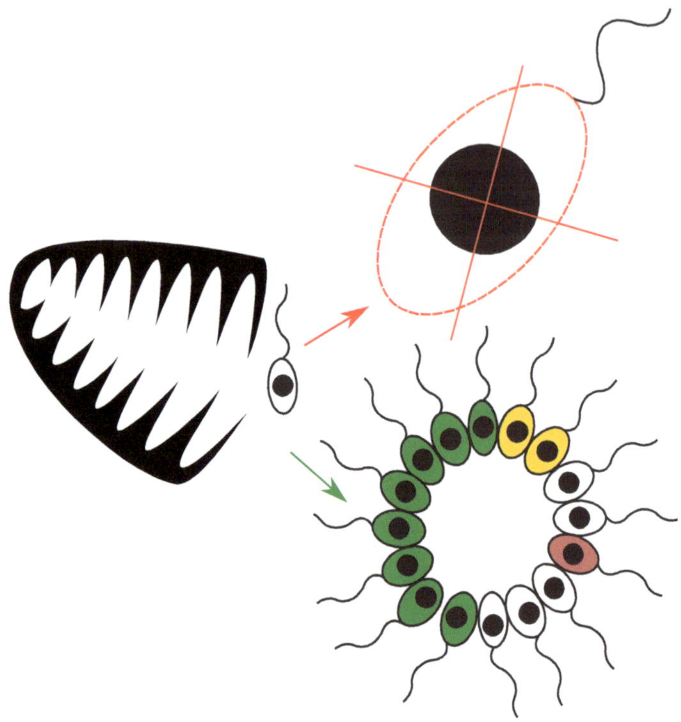

Figure 4.7. Origin of multicellularity. It is not feasible just to enlarge cell, surface is too small. But if cells do not part after mitosis, they might form the body which is big enough to escape from predators. This also provide with new mode of growth and possibility of the division of labor (colored cells).

Multicellularity allows these cells also to divide the labor and cooperate. This is extremely important for the future evolution.

Cells in the multicellular body are not connected forever. Sometimes, one or few cells escape and start a new body. This body will be exact copy (clone) of the previous one (*vegetative reproduction*). It is also possible that when these "escaped cells" go the different route: they become "sex delegates", *gametes*. All gametes want syngamy, and these cells will search for the partner of the same species but with another genotype. In case of heterogamy and oogamy, it is easy to recognize because genders will provide a hint: male will search for the female. In case of isogamy, gametes search for the partner with different surface proteins. After they finally mate, a diploid cell (zygote) appears. Zygote may winter and then divide meiotically. This is the simplest life cycle of multicellular organism (Fig. 4.8), quite similar to the cycle discussed above for unicellular organism.

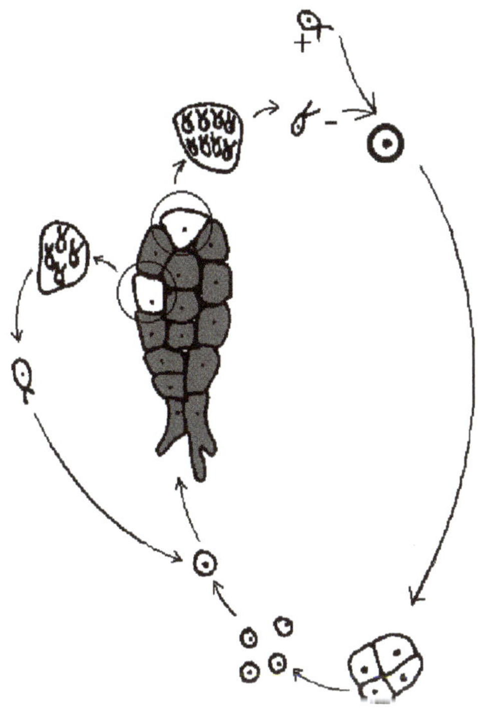

Figure 4.8. Most ancient life cycle of the multicellular organism. Zygote does not grow, it divides meiotically. Somatic ("grey") cells are going to die, only germ cells transfer their DNA to future generations.

However, frequently zygote starts to grow and divide mitotically, making the diploid body. There are two reasons to make multicellular body out of zygote without meiosis: (a) because in can and (b) because diploid is better. "It can" because zygote already contains DNA program about how to build multicellular body. Why diploid is better, explained in next section.

If multicellular organism consists of diploid cells ($2n$), we will use the neutral term **diplont**. Multicellular organisms with haploids cells (n) are **haplonts**.

"Escaped cells", "sex delegates", or mother cells of gametes from the above is a first stage of the division of labor when cells are separating into two types, **germ cells** and **somatic cells**. Somatic cells are those which will eventually die, but germ cells are capable of giving offspring. Having germ cells is not absolutely necessary for multicellular organisms, but most of them have well separated germ lines. Thus, origin of death is directly connected with this separation: somatic cells are not needed for future generations. Unicellular organisms are potentially immortal, and same are cancer cells which also escape from organism (but they cannot make the new one).

Life cycle of multicellular organism could be described starting from haplont (Fig. 4.9). When environment conditions are favorable, it has vegetative reproduction. One variant of vegetative reproduction is that cell (**mitospore**) separates itself from a haplont, then divides into more cells and becomes a new haplont. Sometimes, whole chunks are separated and grow into new haplonts. When conditions change, haplont may start the sexual reproduction: syngamy. In syngamy, one gamete separates from the haplont and unites with a gamete from another haplont. Together, gametes form a zygote. This zygote might go straight to meiosis (as it happens in unicellular eukaryotes) but more frequently, zygote will grow, divide mitotically and finally becomes a diplont. This diplont might be superficially almost identical to haplont but every cell of it contains diploid nucleus (every chromosome has a pair). Diplont (similarly to haplont) may reproduce itself vegetatively (make clones): cell separates itself from a diplont, then divides mitotically into more cells and becomes a new diplont.

The diplont is also capable for asexual reproduction: there could be a cell separates itself from a diplont and divides with *meiosis* creating four spores, each of them will grow into *haplont*.

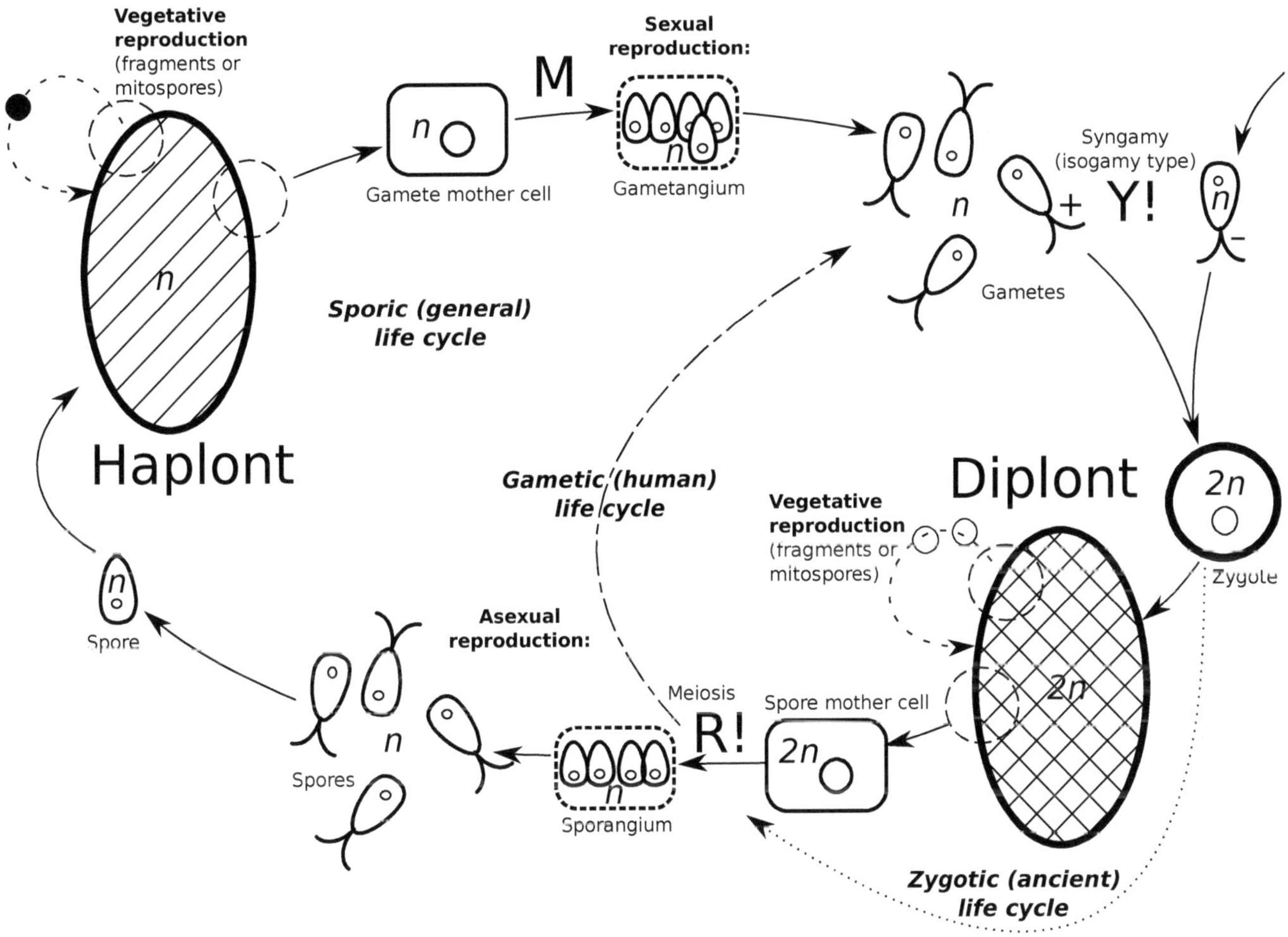

Figure 4.9. General life cycle. Haploid part is on the left, diploid on the right, syngamy on the top, meiosis on the bottom. ”M” letter is used to label mitosis.

4.4.2 Sporic, Zygotic and Gametic Life Cycles

The life cycle described above is the **sporic** life cycle (Fig. 4.10). Organisms with sporic life cycle have both diplont and haplont, equally or unequally developed.

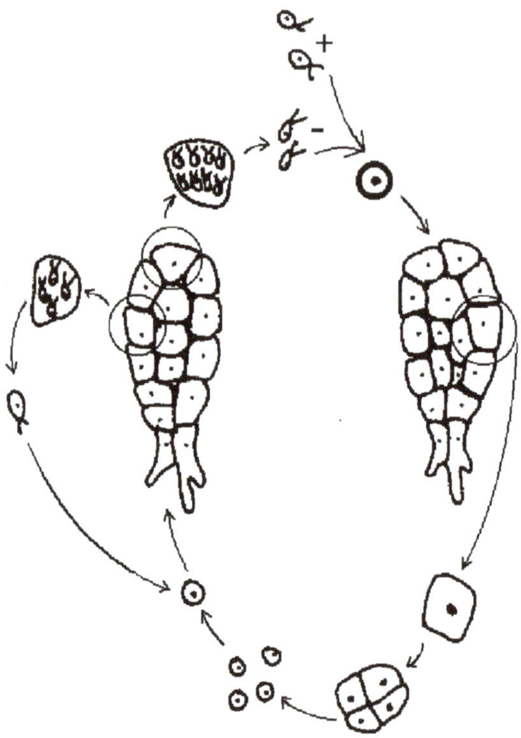

Figure 4.10. Sporic life cycle. Overview. Haploid part is on the left, diploid on the right, syngamy on the top, meiosis on the bottom.

In all, there are three types of life cycles: sporic, zygotic, which is the most similar to unicellular and most primitive; and gametic, which is used by animals and a few protists (Fig. 4.11). The **zygotic** life cycle starts with syngamy and goes to meiosis. It has no diplont. **Gametic** life cycle goes from meiosis to syngamy. It has no haplont.

Protists have all three types of life cycles whereas higher groups have only one. Animals exhibit gametic cycle, whereas plants$_2$ retained the more primitive sporic cycle.

4.4.3 Evolution of Life Cycles

The most striking difference between unicellular and multicellular life cycles is that zygote of multicellular organism may start to make diploid body (diplont) which sometimes is visually almost identical to haplont. This is because in the evolutionary

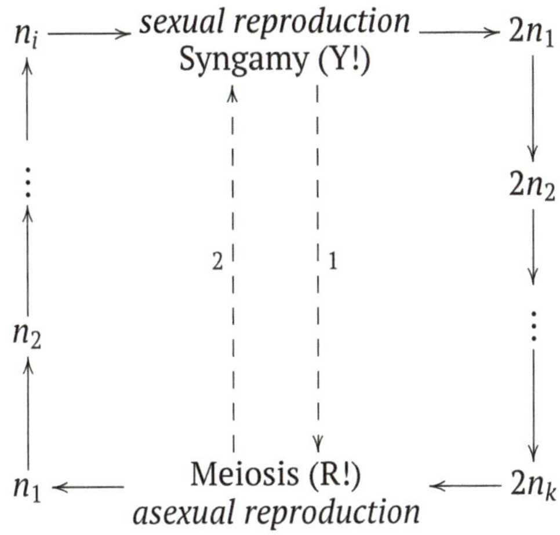

Figure 4.11. Schematic representation of three types of life cycle. The whole scheme is a sporic cycle; 1 zygotic cycle (Y!→R!); 2 gametic cycle (R!→Y!).

perspective, diplonts are "better" than haplonts. Frequent situation of gene dominance allows only one variant (allele) of the gene to work, that may save organism from lethal mutations. An increased number of genes could help to make more proteins. A third reason is that diplonts' genomes are more diverse. One gene may be able to withstand one group of conditions, and the other variant may have a different set of possible conditions. Therefore, diplont is able to take advantage of the capabilities of both genetic variants.

As a consequence, the evolution of life cycles goes from zygotic (similar to unicellular) to the sporic cycle (Fig. 4.12), and then to the more and more expressed domination of diplont, and finally to the complete reduction of haplont, gametic life cycle. It is still an open question how zygotic protists evolved to the sporic side. Most probably, zygote (which is diploid by definition) did not want to divide meiotically. Instead, it grows (which is seen in some protists) and divides mitotically, giving birth to the diplont. This is how first sporic cycle started. The last step of this evolutionary chain was a complete reduction of haplont: after meiosis, spores were replaced with gametes which immediately go to syngamy.

4.4.4 Life Cycle of Vegetabilia

Ancestors of Vegetabilia (plants$_2$) were green algae with zygotic life cycle. It could be imagined that their zygote started to grow because these organisms inhabited shallow waters and want their spores to be distributed with a wind. One way for this to happen is to have the spores on the stalk of the plant. This is probably the reason

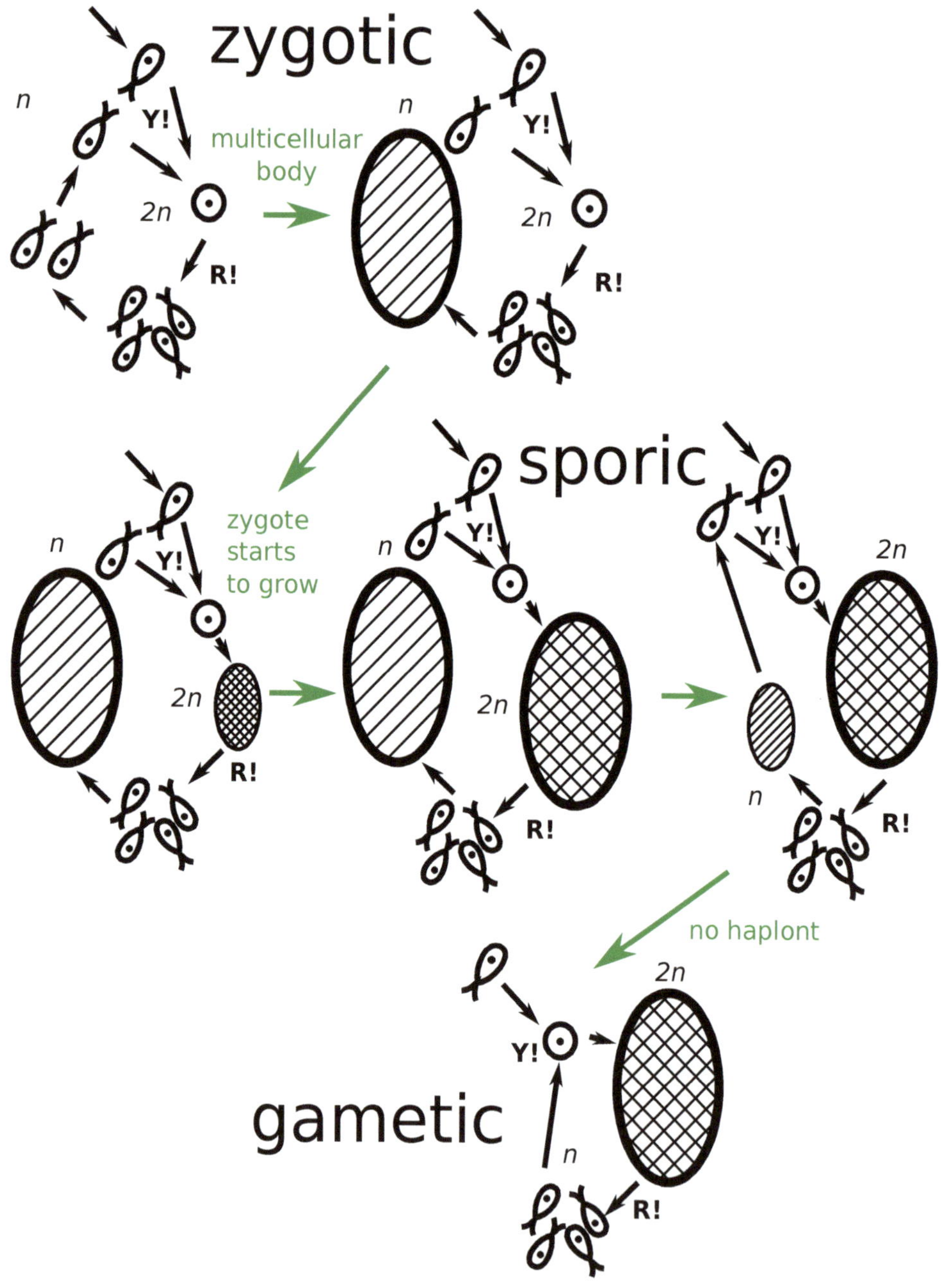

Figure 4.12. The evolution of life cycles (green arrows represent five evolutionary transitions) from unicellular zygotic to multicellular gametic through different variants of sporic cycles.

of zygote growth: primordial diplonts of plants$_2$ were simply **sporangia**, structures bearing spores. Then the benefits of diploid condition described above started to appear, and these primitive plants went onto the road of haplont reduction. However, some Vegetabilia (liverworts, mosses and hornworts), still have haplont domination. This is probably because their haplonts are *poikilohydric* (it is explained in next chapters), adaptation which is beneficial for small plants.

Life cycle of plants$_2$ is sporic, but the science tradition uses plant-related names for the stages. The cycle (Fig. 4.13) begins with a diplont called a **sporophyte**, which produces spores. Sporophyte bears a **sporangium**, inside which **mother cell of spores** uses meiosis to make spores. The spores germinate and grow into haplont called **gametophyte**. Gametophyte produces gametes, specifically a spermatozoa (or simply "sperms") and an oocyte (egg cell). These gametes are developed in special organs—**gametangia**. Gametangium which contains male gametes (sperms) is called **antheridium**, and female gametangium is **archegonium**, the last normally contains only one egg cell (oocyte).

By syngamy (oogamy in this case), the two gametes form a zygote. Next, a *young sporophyte grows on the gametophyte*, and finally, the cycle starts again. Again, sporophyte of Vegetabilia starts its life as a parasite on gametophyte. Even flowering plants have this stage called **embryo**. Maybe, this is why the gametophyte of plants$_2$ has never been reduced completely to transform their cycle into gametic. Even in most advanced plant lineages, their male (which makes only sperms) and female gametophytes have minimum 3 and 4 cells, respectively, but not 0!

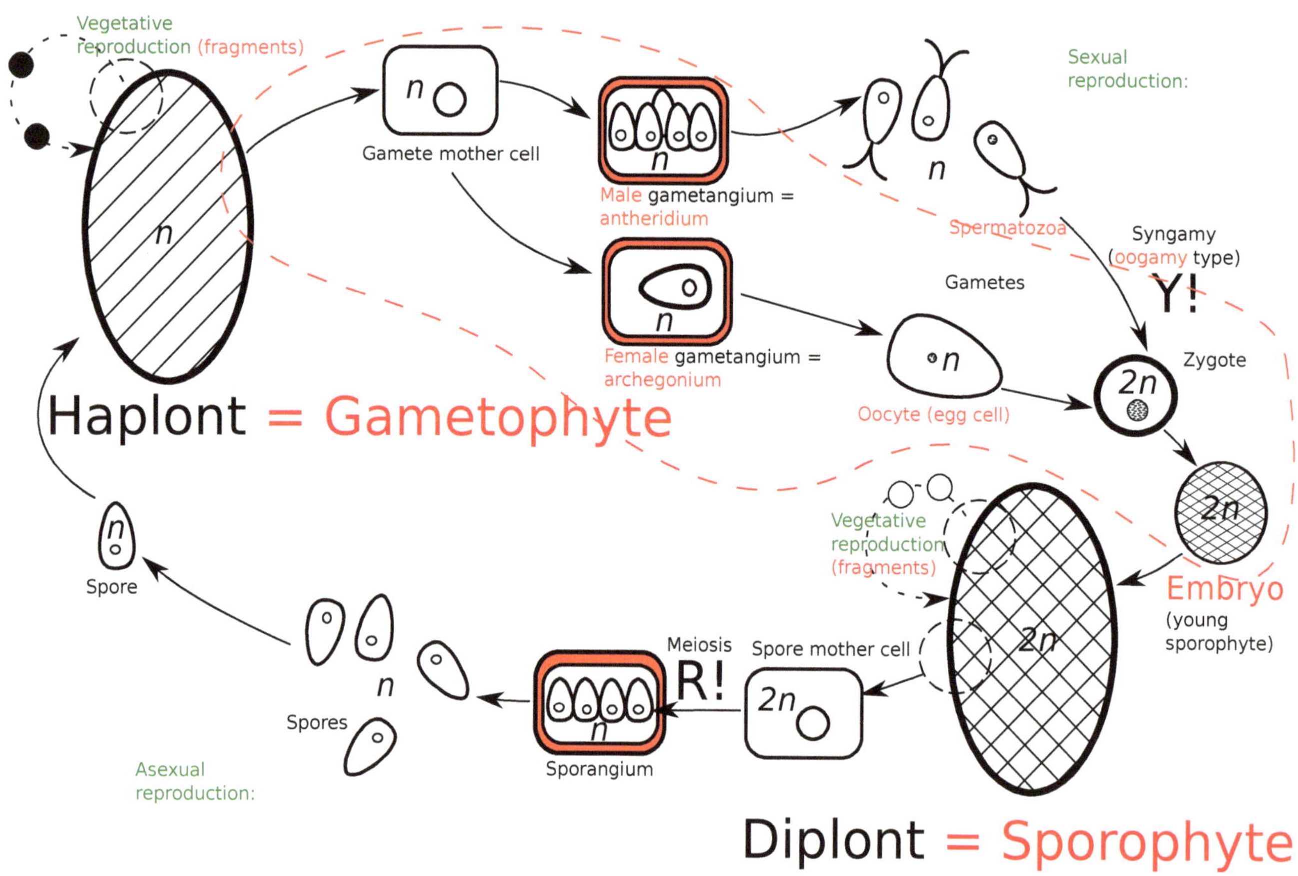

Figure 4.13. Life cycle of land plants. Red color is used for innovations, comparing with previous (general) life cycle scheme.

Chapter 5

Tissues and Organs; or how the Plant is built

5.1 Tissues

From now on, we will frequently use multiple names of plants$_2$ group, they are summarized on Figure 5.1, and in more details—on Fig.6.1.

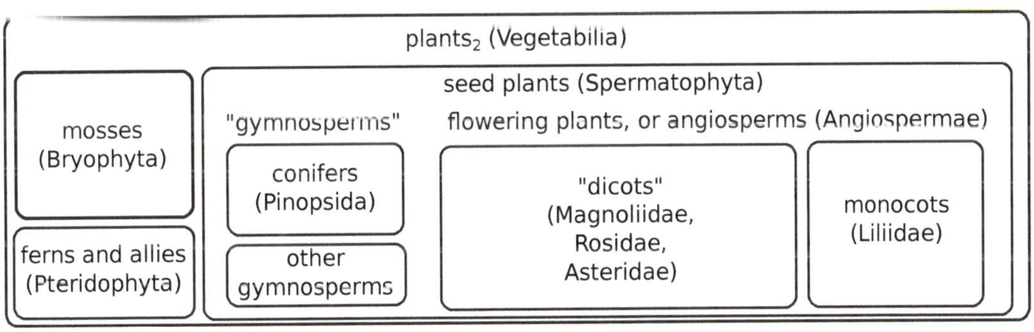

Figure 5.1. Plants$_2$ classification: overview.

5.1.1 Epidermis and Parenchyma

Why did plants go on land? In order to escape competition with other plants for resources like the sun and nutrients, but also to obtain much more sunlight that was otherwise seriously reduced underwater. The move to land also helped plants escape predators. Lastly, plants benefited from this change because they escaped from the **temperature-gases conflict**: warmer temperatures are good for organisms but significantly decrease the amount of gases diluted in water.

Although this action solved several problems, it also raised new issues that needed to be dealt with. The most important was the risk of drying out. To combat this, plants developed their first **tissue**: **epidermis** covered with a **cuticle** which served a purpose similar to a plastic bag. For the really small (millimeters) plant it is enough because, in accordance to **surface / volume law**, they have high relative surface, and diffusion can serve for gas exchange.

However, bigger plants also need to exchange gases, and they developed **stomata** which served as a regulated pore system. The remaining cells became second tissue: **parenchyma** (or **ground tissue**, or **main tissue**).

Another response (Fig. 5.3) for drying was a development of *poikilohydricity* (see below), the ability to hibernate in (almost) dried condition. As hibernation is generally dangerous since it requires "system restart", that evolutionary route did not become the main.

Tissue is a union of cells which have common origin, function and similar morphology. Tissues belong to organs: **organ** is a union of different tissues which have common function(s) and origin. Plants have simple and complex tissues. The **simple tissues** are composed of the same type of cells; **complex tissues** are composed of more than one type of cell, these are unique to plants.

Parenchyma (Fig. 5.4) are spherical, elongated cells with a thin primary cell wall. It is a main component of young plant organs. The basic functions of parenchyma are *photosynthesis and storage*. Parenchyma cells are widespread in plant body. They fill the leaf, frequent in stem cortex and pith and is a component of complex vascular tissues (see below). Contrary to parenchyma (which is a simple tissue), **epidermis** is a complex tissue composed of epidermal and stomata cells. Its main functions are *transpiration, gas exchange and defense*.

As it seen here, plants acquired tissues in a way radically different from animals(Fig. 5.2) : while plants regulate gas and water exchange in response to terrestrial environment, animals actively hunt for food (using kinoblast tissues) and then digest it (with pagocytoblast tissue).

5.1.2 Supportive Tissues: Building Skyscrapers

When more and more plants began to move from the water to the land, competition once again became a problem (Fig. 5.3). To solve this, plants followed "Manhattan solution": they grew vertically in order to be able to escape competition for the sunlight and therefore must develop **supportive tissues**.

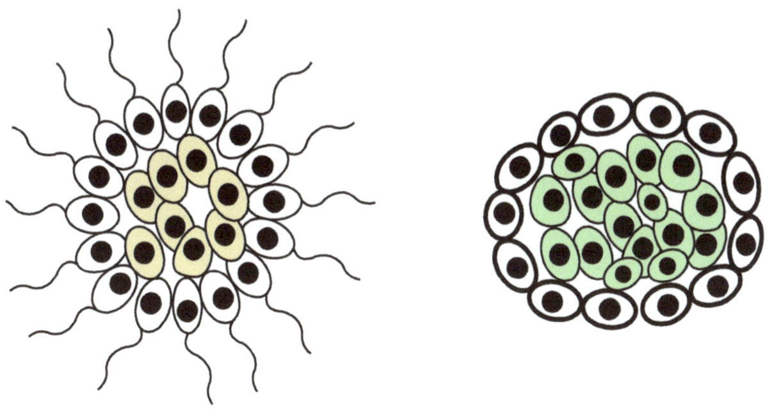

Figure 5.2. Phagocytella (proto-animal) with kinoblast and phagocytoblast *vs.* proto-plant with epidermis and ground tissue.

Collenchyma (Fig. 5.4) is living supportive tissue that has elongated cells and a thick primary cell wall. Its main function is the mechanical support of young stems and leaves via turgor.

Sclerenchyma (Fig. 5.4) is a dead supportive tissue that consists of long fibers or short, crystal-like cells. Each cell has a thick *secondary wall* that is rich in lignin. Its main function is a support of older plant organs, and also hardening different parts of plants (for example, make fruit inedible before ripeness so no one will take the fruit before seeds are ready to be distributed). Without sclerenchyma, if a plant isn't watered, the leaves will droop because the vacuoles will decrease in size which lowers the turgor. Fibers inside phloem (see below) are sometimes regarded as a separate sclerenchyma.

Three times in their evolution plants found the new application for lignin or similar polymers: at first, similar chemicals covered the spore wall which was an adaptation to the spore distribution with wind. Then similar chemicals were used to make cuticle, "epidermal plastic bag" to prevent transpiration outside of stomata. Finally, with acquiring of sclerenchyma, plants found how to use dead cells with completely lignified cell walls.

By the way, stomata likely had a similar fate, they historically appeared on sporangia to help them dry faster and release spores effectively. Regulation of transpiration is their second function.

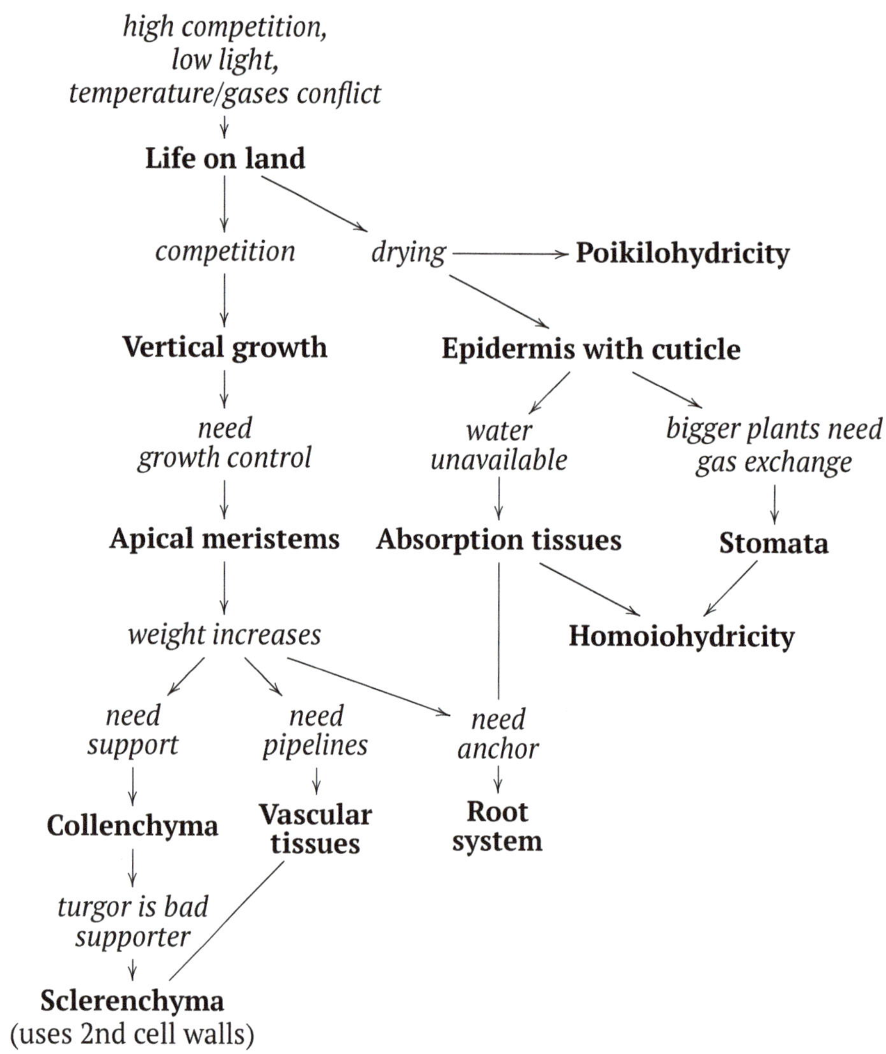

Figure 5.3. *Challenges* to land plants and their **responses**, part 1. Part 2 is on Fig. 7.14.

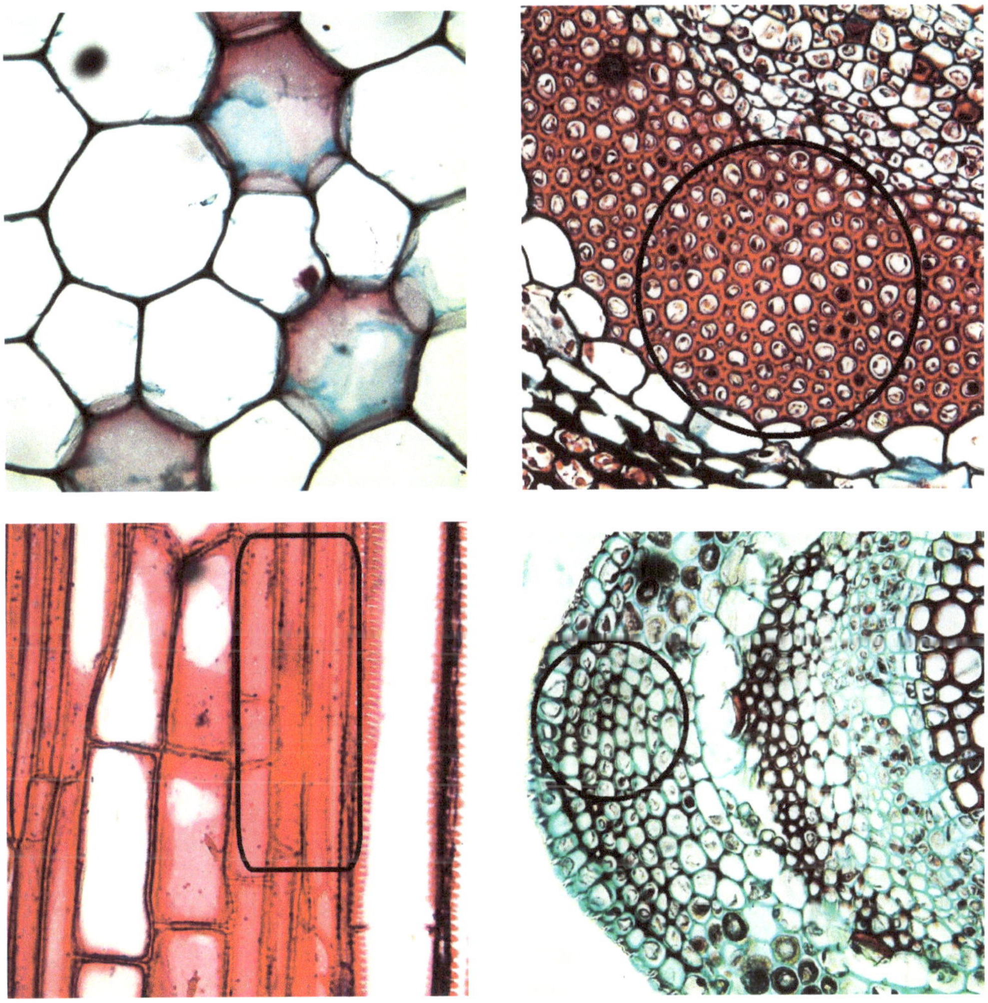

Figure 5.4. Left to right, top to bottom: parenchyma, sclerenchyma (cross- and longitudinal sections) and collenchyma. First three photos from the stem of *Helianthus*, fourth from *Medicago* stem. Magnification $\times 400$.

Cell types and tissues

"Parenchyma" and "sclerenchyma" terms are freShoot systemquently used in two ways: first, to name tissues (or even classes of tissues) which occur in multiple places of the plant body, and second, to name the cell types which are components of tissues. Therefore, it is possible to say "parenchyma of stem", "parenchyma of stem pith", "parenchyma of xylem" and even "leaf mesophyll is a parenchyma".

5.1.3 Meristems: the Construction Sites

Plant growth requires centers of development which are **meristems**. **Apical meristems** are centers of plant development located on the very ends of roots (**RAM**) and stems (**SAM**). They produce intermediate meristems (like **procambium**) which form all **primary tissues**. The **lateral meristem** or **cambium** originates from the procambium which in turn originates from apical meristems. It usually arises between two vascular tissues and its main functions are thickening and producing **secondary vascular tissues** (Fig. 5.5).

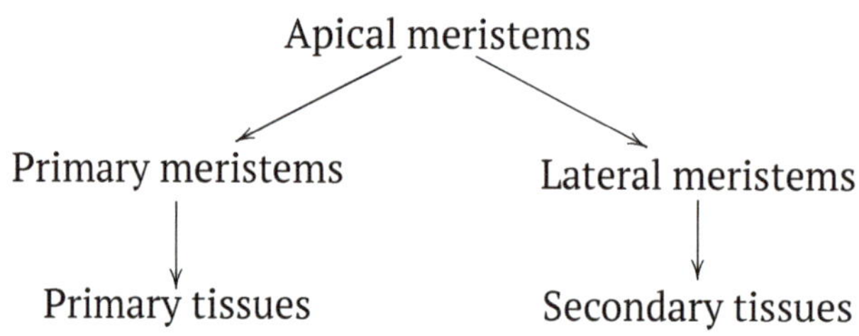

Figure 5.5. Meristems and tissues.

Other meristems include: **intercalary** which elongate stems from the "middle", **marginal** which are responsible for leaf development and **repair** meristems arising around wounds, they also control vegetative reproduction.

5.1.4 Vascular Tissues

Bigger plants escaped from competition and performed effective metabolism. However, with all the growth the plants went through, their size became too big for slow symplastic plasmodesmata connections. Another, filter paper-like apoplastic transport was also not powerful enough. The solution was to develop **vascular tissues**, xylem and phloem (Fig. 5.6, Fig. 5.29).

The main functions of xylem are the transportation of water and mechanical support. The **xylem** may be found either in a vascular bundle or a vascular cylinder. The three types of xylem cells are **tracheary elements** (these include **tracheids** and

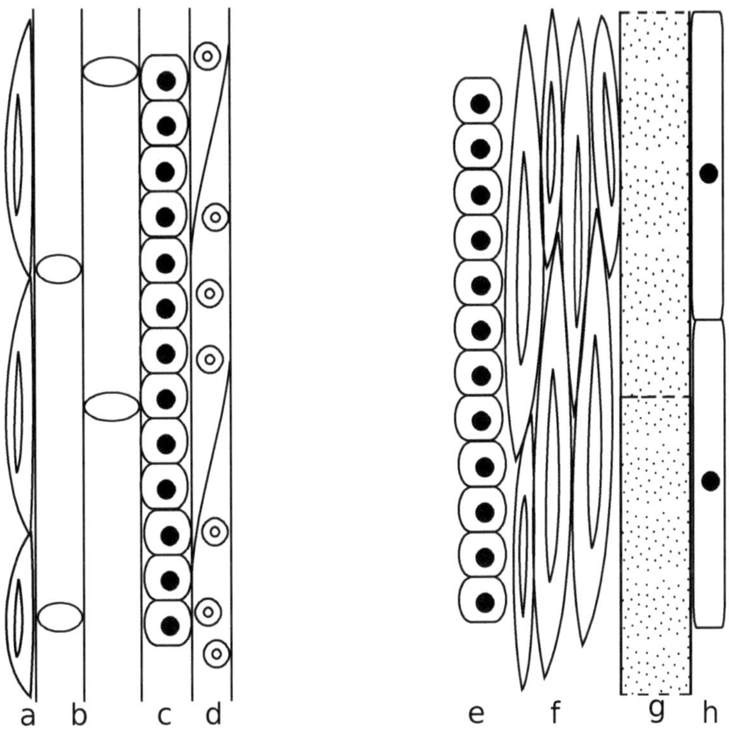

Figure 5.6. Cells of xylem (left, a–d) and phloem (right, e–h): a fibers, b vessels with open perforations, c parenchyma, d tracheids with pits, e parenchyma, f fibers, g sieve tubes, h companion cells.

vessel members), **fibers**, and **parenchyma**. Xylem elements, except for the parenchyma, are rich in lignin and are main components of wood. Tracheids are closed on both ends and connected with **pits** whereas vessel members are more or less open and connects via **perforations**. Tracheids, vessel members and fibers are dead cells. Xylem parenchyma, on the other hand, is alive.

Pits of tracheids consist of a pit membrane and the torus in a center, there are *no openings*. The presence of tracheids and/or vessel elements has evolutionary significance. Vessels (made of vessel members) are more effective; consequently, more "primitive" plants have more tracheids whereas more "advanced" have more vessel members. As an example, gymnosperms have only tracheids while most flowering plants have tracheids and vessel members. Individual development also mimics this evolutionary trend. Younger flowering plants have more tracheids whereas mature plants have more vessel members. Primary xylem mostly has tracheids and vessels with scalariform perforations whereas secondary xylem (which originates from cambium) consists mostly of vessels with open perforations. The common name for secondary xylem is **wood**.

It is a mistake to think that tracheids are better than vessels. In fact, the main problem is frequently not too slow but too fast water transport. Tracheids have an advanced connection system (called torus) which has the ability to close pore if the water pressure is too high and therefore more controllable. Leaking would be less dangerous in tracheids. And in water-poor environments (like taiga in winter), plants with tracheids will have the advantage. Contrary, having vessels is like to have race car for ordinary life; only flowering plants "learned" how to use them effectively.

Dead cells are useful but hard to control. However, if xylem transport needs to be decreased, there is a way. Xylem parenchyma cells will make **tyloses** ("stoppers") which will grow into dead tracheary elements and stop water if needed. Many broadleaved trees use tyloses to lower xylem transport before the winter.

The **phloem** generally occurs adjacent, or right next to, the xylem, with the xylem facing the inner part of the plant and the phloem facing the outer part of the plant. The main functions of the phloem are the transportation of sugars and mechanical support. The four types of phloem cells are: **sieve tube cells**, **companion cells**, **fibers** (the only dead cells in phloem), and **parenchyma**. Sieve tube cells of flowering plants have cytoplasm flowing through perforations (sieve plates) between cells but do not contain nuclei. Companion cells will make proteins for them. However, in gymnosperms and more "primitive" plants there are no companion cells at all, so sieve tube cells do contain nuclei. This is comparable to red blood cells in vertebrates: while mammals have them anucleate, erythrocytes of other vertebrates contain nucleus. The secondary phloem generally has more fibers than the primary phloem.

This small table summarizes differences between xylem and phloem:

	Xylem	**Phloem**
Contains mostly	Dead cells	Living cells
Transports	Water	Sugar
Direction	Up	Down
Biomass	Big	Small

5.1.5 Periderm

Periderm is a secondary dermal tissue which arises inside the stem ground tissue, closer to the surface. Like the other dermal tissue (epidermis), it is a complex tissue. It includes three layers (starting from surface): **phellem** (**cork**), **phellogen** (**cork cambium**) and **phelloderm** (Fig. 5.7). Phellem consists of large dead cells with secondary walls saturated with suberin, and is the main, thickest component of periderm. Phellogen is a lateral meristem, like cambium; it often arises fragmentarily (and also temporarily) and does not cover the whole stem under-surface. But when phellem starts to grow, all peripheral tissues (like epidermis) will be separated from water transport and eventually die. Phellogen makes phellem towards the surface, and phelloderm towards the next layer (phloem). Phelloderm is a minute tissue, and does not play significant role in the periderm.

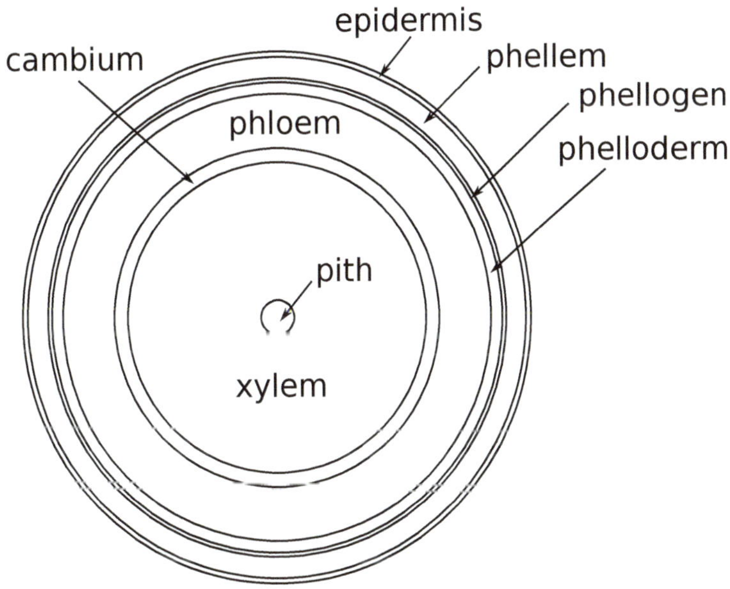

Figure 5.7. Principal location of stem tissues (simplified).

In older plants, phellogen arises deeper, sometimes inside phloem and separates outer layers of phloem from vascular cylinder. All this mixture of tissues (phellogen, phellem, phelloderm, epidermis and upper layers of phloem) considered as a **bark**.

5.1.6 Absorption Tissues

Poikilohydric plants do not save water and they can survive even complete desiccation because their cells will hibernate. An example of a poikilohydric plants would be mosses. **Homoiohydric** plants (which are majority of plants$_2$), however, do save water. They try to support the water content and do not survive complete desiccation. An example of a homoiohydric plant would be any "typical" plant, saying, corn. Somehow similar traits are comparable in poikilothermic animals, such as reptiles, and homoiothermic animals, such as birds and mammals, except in reference to body heat rather than water conservation.

Absorption tissues are always simple, primary tissues. Most important of them is **rhizodermis** (rhizoderm), or **root hairs**, which originates from protoderm (protoepidermis), but its lifespan is much shorter than of epidermis. There are other absorption tissues, for example, **velamen**, which originates from the root cortex and consists of large, empty, easy to get wet dead cells.

5.1.7 Other Tissues

Secretory tissues spread across the plant body, concentrating in leaves and young stems. These tissues may secrete latex, volatile oils, mucus and other chemicals. Its functions can be attraction or dis-attraction, communication or defense, and many others.

In addition to tissues, plant body may contain **idioblasts**, cells which are quite dissimilar from surrounding cells. Idioblasts used for accumulation of unusual (and possibly dangerous) compounds like **myrosinase**, protein splitting glucosinolates into sugars and toxic isothiocyanate (**mustard oil**). We use mustard oil as a spice but for the plant, it works like a binary chemical weapon against insect herbivores: when myrosinase-containing idioblasts are damaged, mustard oil kills damaging insects. Among plants, the whole order Brassicales from rosids is capable to produce myrosinase, examples are different cabbages (*Brassica* spp.), papaya (*Carica*), horseradish tree (*Moringa*) and many others.

5.2 Organs and Organ Systems

Vegetabilia (Fig. 6.1) have three different types of body construction (Fig. 5.8). The most primitive plants have thallus body, more advanced is the shoot (unipolar) plant body, and most land plants have the bipolar plant body. The **thallus** plant body is flat, similar to leaf but do not differentiated into particular organs. Most gametophytes (except true mosses) have this type, and also few sporophytes (which mostly are reduced water plants). **Shoot (unipolar) plant body** consists only of branching shoots, roots are absent. This is typical to all Bryophyta sporophytes, mosses (Bryopsida) gametophytes, and also to sporophytes of Psilotopsida (whisk ferns). Finally, **bipolar plant body** has both shoots and roots (Fig 5.11). Most bipolar plants have shoots consist of stems and leaves, but this is not an absolute requirement since young plant stems are normally green and can do photosynthesis.

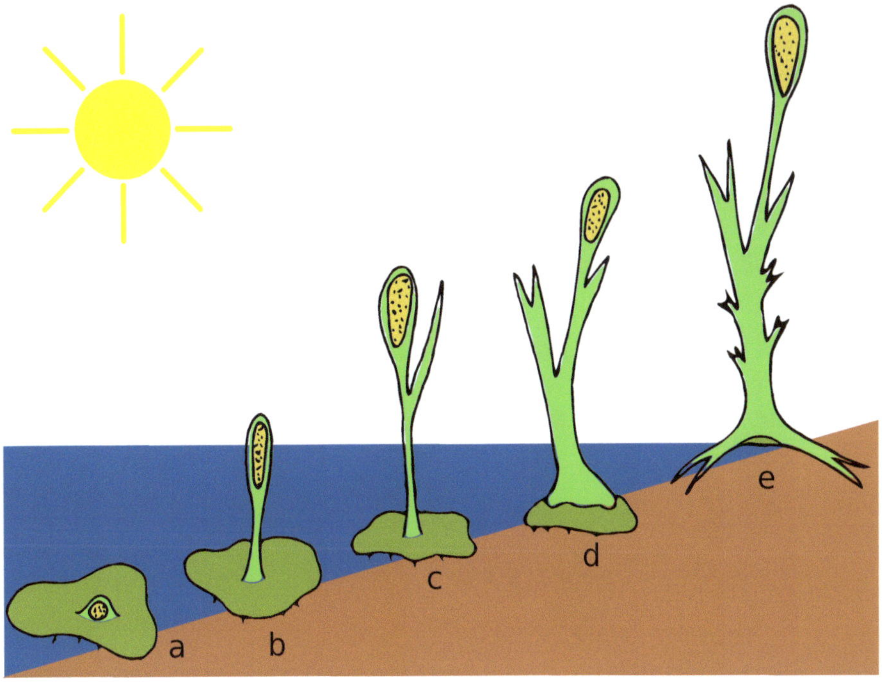

Figure 5.8. Evolution of $plants_2$ body types: a–e thallus gametophytes, a thallus sporophyte, b–d shoot sporophyte, e bipolar sporophyte.

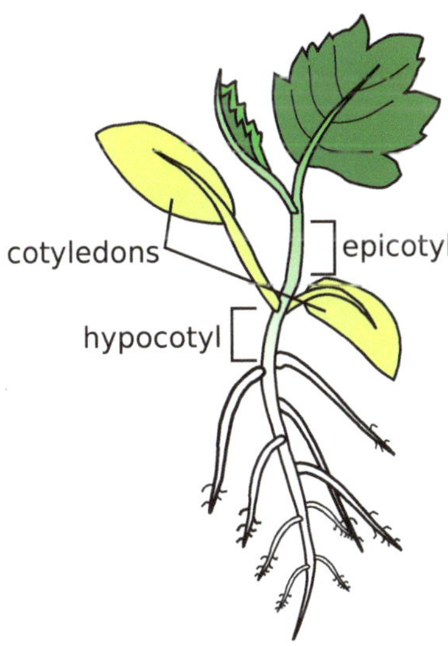

Figure 5.9. Young seedling with epicotyl and hypocotyl.

Typical organs of bipolar plant are stems (axial aerial organs with continuous growth), leaves (flat lateral organ with restricted growth), roots (axial soil organ modified for absorption) and **floral units (FU)** which are elements of the generative system (fructifications) such as a pine cone or any flower.

Buds, fruits, seeds and specific to seedlings hypocotyl and epicotyl are **non-organs** for different reasons: **buds** are just young shoos, **fruit** is the ripe flower, **hypocotyl** is a part of stem between first leaves of the seedling (cotyledons) and root (i.e., stem/root transition place), **epicotyl** is first internode of stem (Fig. 5.9), and finally, **seed** is a chimeric structure with three genotypes so it is impossible to call it "organ".

Root, stem, leaf and FU are *four basic plant organs* (Fig. 5.10) which in bipolar plant could be grouped in root and shoot system; the latter is frequently split into *generative shoot system* (bearing FU), and *vegetative shoot system* (without FU).

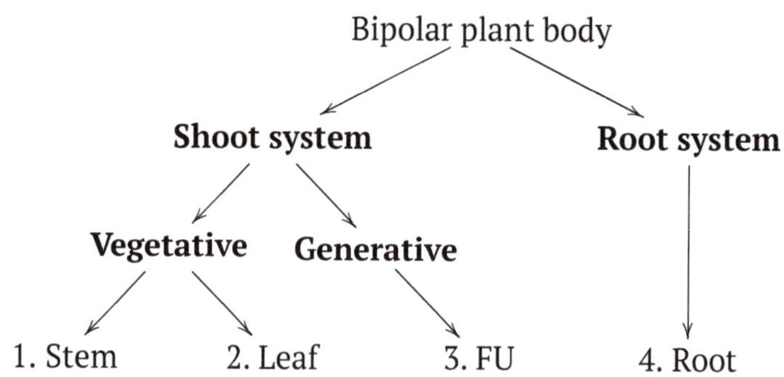

Figure 5.10. Bipolar plant: **organ systems** and four organs.

Vegetative shoot system usually consists of main and secondary shoots; shoots contain terminal buds, axillary (lateral) buds, stem (nodes and internodes) and leaves. We will start from leaves.

5.3 The Leaf

The first and ultimate goal of every plant is photosythesis. If a plant is multicellular, it usually develops relatively large, flat structures which goal is to catch sun rays. Terrestrial plants are no exception; most probably, they started to build their body with organs similar to present day leaves.

A **leaf** is *lateral photosynthetic organ of shoot with restricted growth.* Its functions are photosynthesis, respiration, transpiration, and synthesis of secondary chemicals. Features of a leaf (i.e., characters help to distinguish it) include having a bud in the

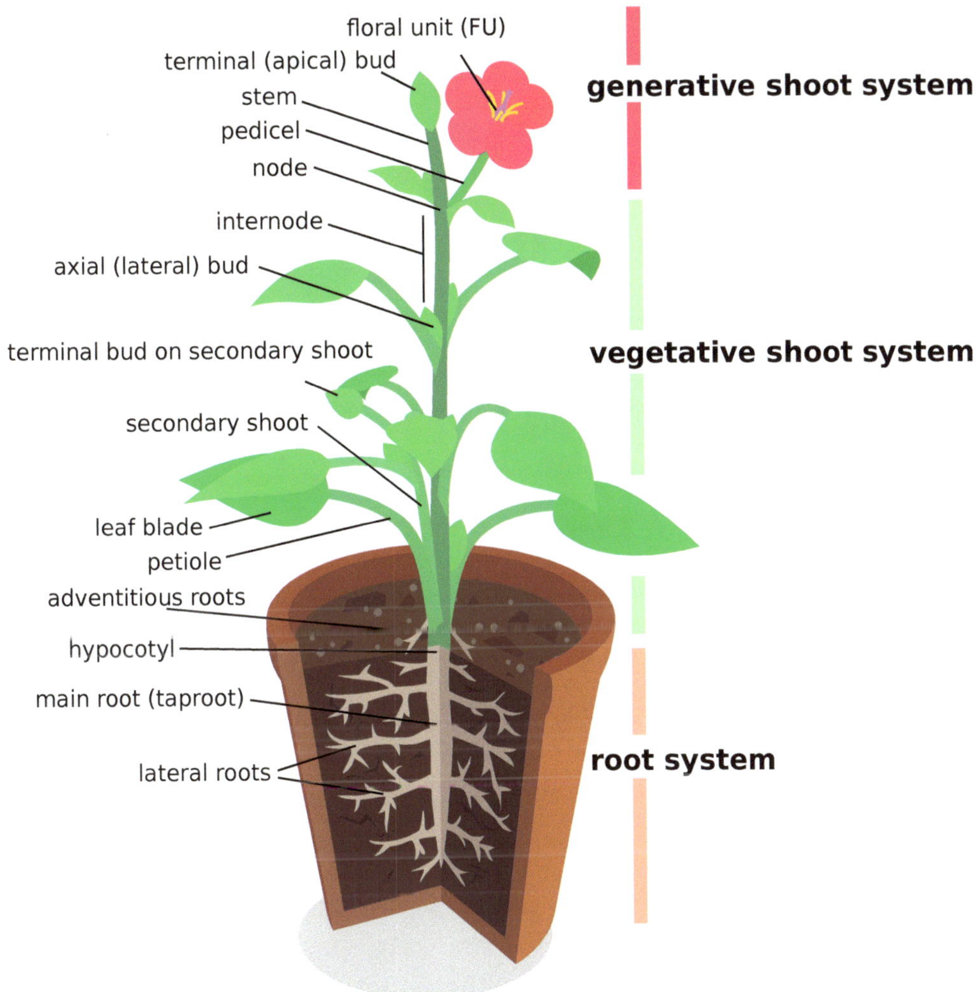

Figure 5.11. Systems of organs and organs of bipolar plant.

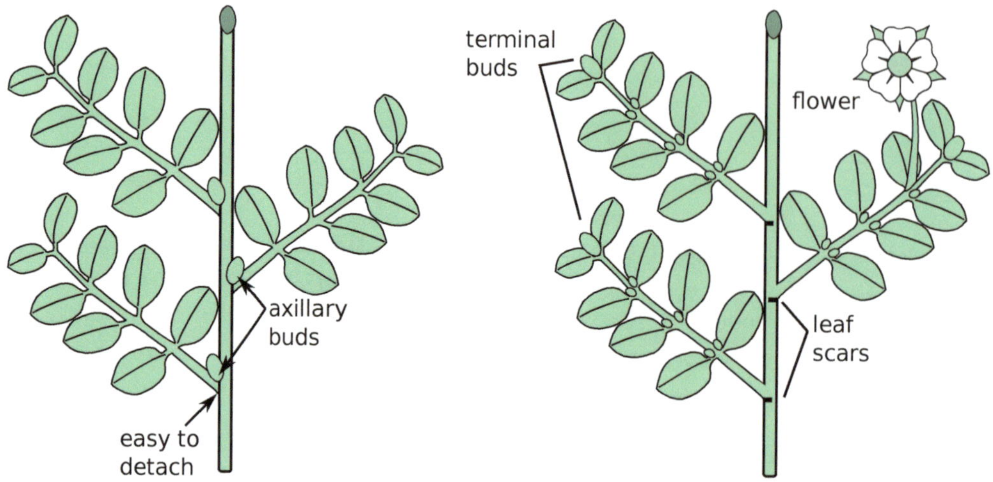

Figure 5.12. How to distinguish compound leaves (left) from branches (right).

axil, not growing by apex, not producing new leaves or shoots, and having *hierarchal morphology* (see below).

5.3.1 Morphology of the Leaf

Morphology means external, well visible structural features whereas **anatomy** needs tools like a microscope and/or scalpel. Leaves are very important in plant morphology. The ability to describe the leaf is a must even for novices in botany.

In all, *plants are fractal organisms*, like Sierpinski triange (Fig. 5.13). All fractals are self-similar (Fig. 5.14), and plants are no exception. Self-similarity, or "Russian doll effect" means that almost every part of plant may be a part of the bigger complex, this bigger one—the part of even bigger system, and so on. This is what we see in leaves as *levels of hierarchy*. **Simple leaves** have just one level of hierarchy whereas compound leaves have two or more levels of hierarchy. **Compound leaves** are sometimes mixed with branches but there are many other characteristics which allow to distinguish them (Fig. 5.12).

Figure 5.13. One of simple fractals: Sierpinski triangle.

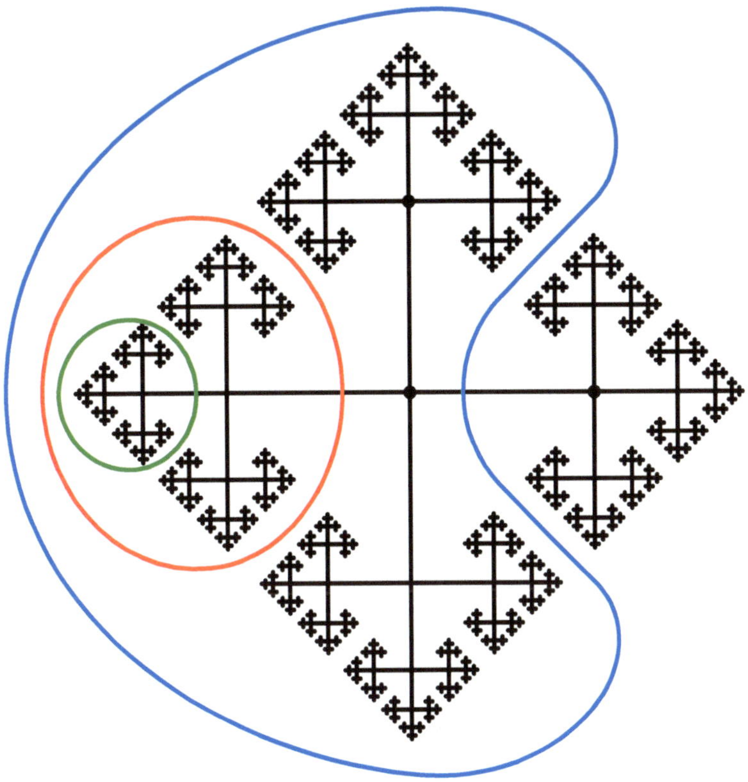

Figure 5.14. The example of self-similarity.

To describe leaves, one should always note the level of hierarchy like "on the first level of hierarchy, the shape is ..., on the second level of hierarchy, the shape is ..." As it was mentioned above, leaf hierarchy is similar to Russian dolls: every smaller doll has a bigger doll (next hierarchy level) outside. For example, if the leaf is compound (consists of multiple leaflets), the overall shape of it could be, saying, round (circular) but the shape of individual leaflet of the very same leaf could be ovate (Fig. 5.15). As a result, the description will say that on first level of hierarchy the leaf is ovate, and on the third level—circular.

There are three types of leaf characters: general, terminal, and repetitive. **General characters** are only applicable to the whole leaf. **Terminal characters** are only applicable to the terminal leaflets. Terminals are the end parts of leaves, they do not split in smaller terminals; clover leaf, for example, has 3 terminals. Lastly, repetitive characters repeat on each level of leaf hierarchy. General and terminal characters do not depend on hierarchy. **Repetitive characters** may be different on each step of hierarchy.

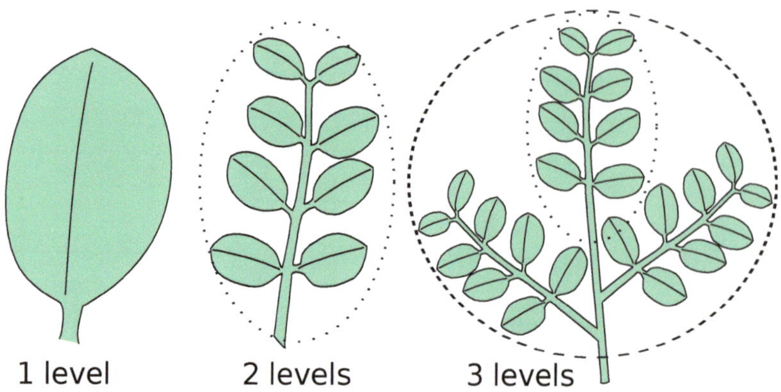

Figure 5.15. Leaves with one, two and three levels of hierarchy. Please note that the last leaf is ovate on the first and second level but circular on the third level of hierarchy.

General characters of leaf include **stipules** and other structures located near leaf base (Fig. 5.16): **sheath** (typical for grasses and other liliids) and **ocrea** (typical for buckwheat family, Polygonaceae).

Figure 5.16. From left to right: sheath, stipules and ocrea.

Repetitive characters are the shape of the leaf (Fig. 5.17), leaf dissection, and whether the blade is stalked (has petiole) or not.

Terminal characters are applicable only to terminal leaflets of leaves. These characters (Fig. 5.19) are the shape of the leaf blade base, the leaf tip, the type of margin, the surface, and the venation. The base of the leaf blade could be **rounded**, **truncate** (straight), **cuneate**, and **cordate**. The leaf apex could be **rounded, mucronate**, **acute**, **obtuse**, and **acuminate**. Leaf margin variants are **entire** (smooth) and **toothed**: **dentate**, **serrate**, **double serrate** and **crenate**.

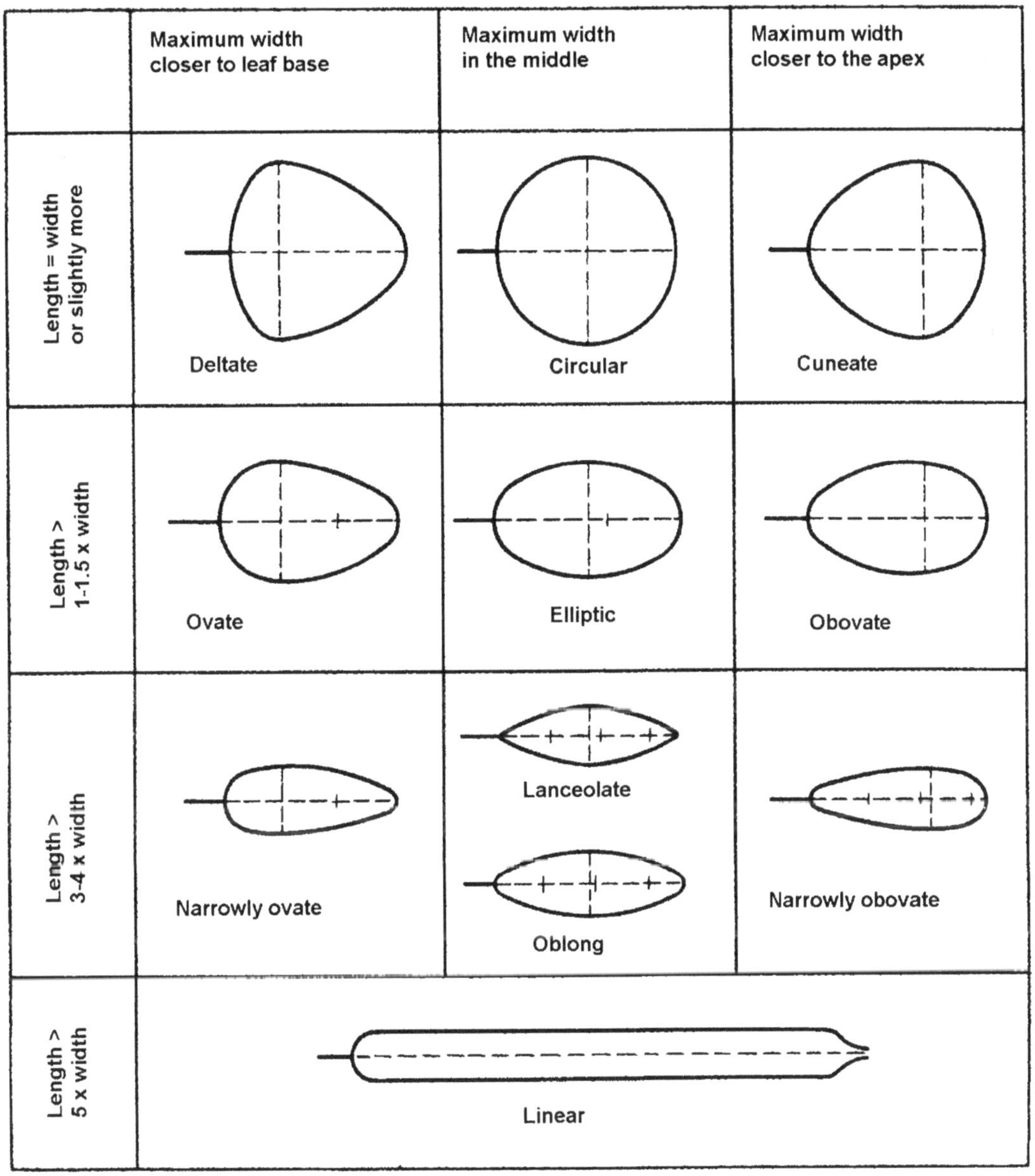

Figure 5.17. Leaf shapes.

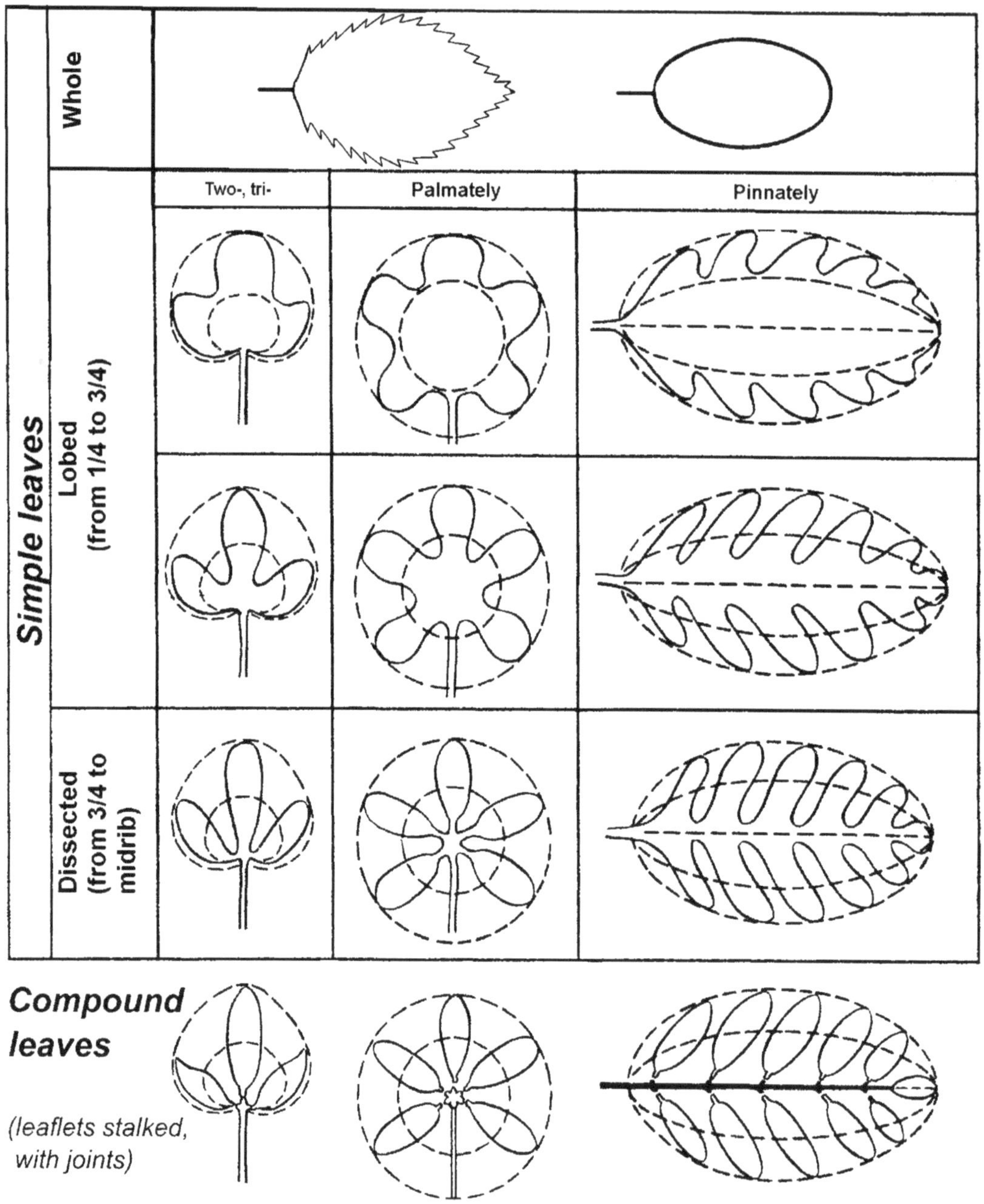

Figure 5.18. Leaf dissection.

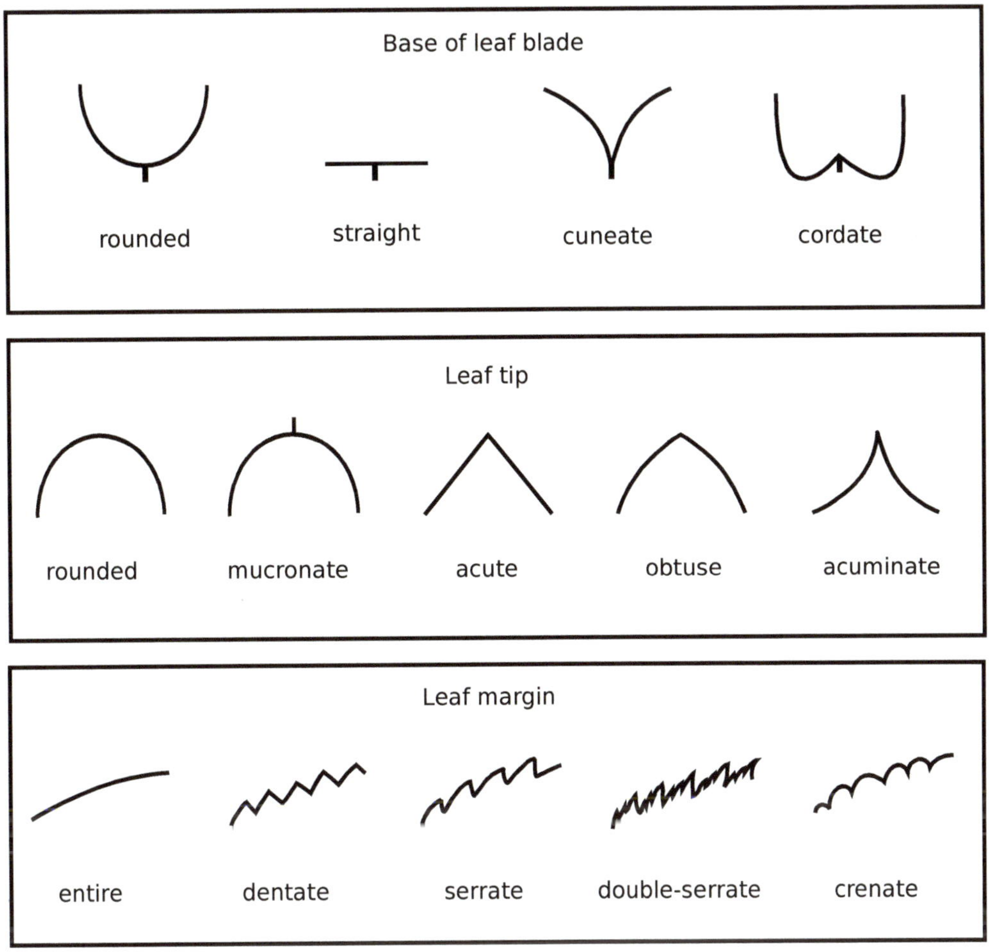

Figure 5.19. Terminal leaf characters.

Leaf veins are vascular bundles coming to the leaf from stem. Frequently, there is a **main vein** and **lateral veins** (veins of second order). There are multiple classifications of leaf venation; and example is shown on the Figure 5.20.

Note that in **dichotomous** venation, each vein divides into two similar parts which is known as dichotomous branching. The example of dichotomous venation is the leaf of maidenhair tree, ginkgo (*Ginkgo biloba*). Another frequently segregated type of venation is **parallellodromous**, but in essence, this is acrodromous venation in linear leaves (for example, leaves of grasses) where most of veins are almost parallel.

To characterize the whole leaf, one might use the following plan:

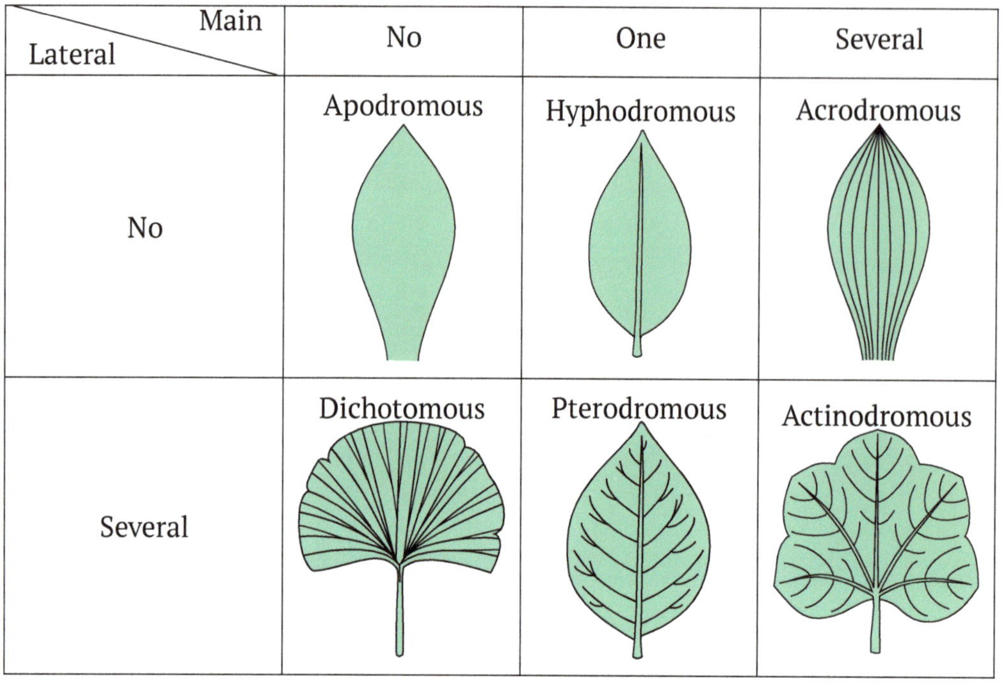

Figure 5.20. The simple classification of leaf venation.

1. General characters (leaf as a whole):
    (a) stipules (present / absent, deciduous / not, how many, size, shape);
    (b) base (sheath / no sheath, ocrea / no ocrea)
2. First level of hierarchy: repetitive characters:
    (a) symmetry (symmetrical / asymmetrical);
    (b) shape;
    (c) dissection;
    (d) petiole (presence and length)
3. Second level of hierarchy
4. Third level of hierarchy, and so on
5. Terminal characters (leaflets):
    (a) base of leaf blade (rounded, truncate, cuneate, cordate);

(b) apex (rounded, mucronate, acute, obtuse, acuminate);

(c) margin (whole, dentate, serrate, double serrate, crenate);

(d) surfaces (color, hairs etc.);

(e) venation (apo-, hypho-, acro-, ptero-, actinodromous)

* * *

Heterophylly refers to a plant having more than one kind of leaf. A plant can have both juvenile leaves and adult leaves, water leaves and air leaves, or sun leaves and shade leaves. A leaf mosaics refers to the distribution of leaves in a single plane perpendicular to light rays, this provides the least amount of shading for each leaf.

Leaves have seasonal lives; they arise from the SAM through **leaf primordia**, and grow via marginal meristems. The old leaves separate from the plant with an **abscission zone**.

* * *

The famous poet and writer Johann Wolfgang Goethe is also considered a founder of plant morphology. He is invented an idea of a "primordial plant" which he called "Urpflanze" where all organs were **modifications** of several primordial ones. In accordance to Goethe's ideas, plant morphology considers that many visible plant parts are just modifications of basic plant organs.

Modifications of the leaf include **spines** or **scales** for defense, **tendrils** for support, **traps**, "sticky tapes", or **urns** for interactions (in that case, catching insects), **plantlets** for expansion, and **succulent** leaves for storage. Plantlets are little mini plants that grow on the main plant and then fall off and grow into new plants; the most known example is *Kalanchoë* ("mother of thousands") which frequently uses plantlets to reproduce. Plants that have insect traps of various kinds are called carnivorous plants (in fact, they are still photoautotrophs and use insect bodied only as fertilizer). Several types of these are the cobra lily (*Darlingtonia*), various pitcher plants (*Nepenthes*, *Cephalotus*, *Sarracenia*), the butterwort (*Utricularia*), the sundew (*Drosera*), and the best known, the Venus flytrap (*Dionaea*).

5.3.2 Anatomy of the Leaf

Anatomically, leaves consist of epidermis with stomata, **mesophyll** (kind of parenchyma) and vascular bundles, or veins (Fig. 5.22). The mesophyll, in turn, has palisade and spongy variants. Palisade mesophyll is located in the upper layer and serves to decrease the intensity of sunlight for the spongy mesophyll, and also catches slanted sun rays. The **palisade mesophyll** consists of long, thin, tightly arranged cells with chloroplasts mostly along the sides. The **spongy mesophyll** cells are roughly packed, they are rounded and have multiple chloroplasts (Fig. 5.21).

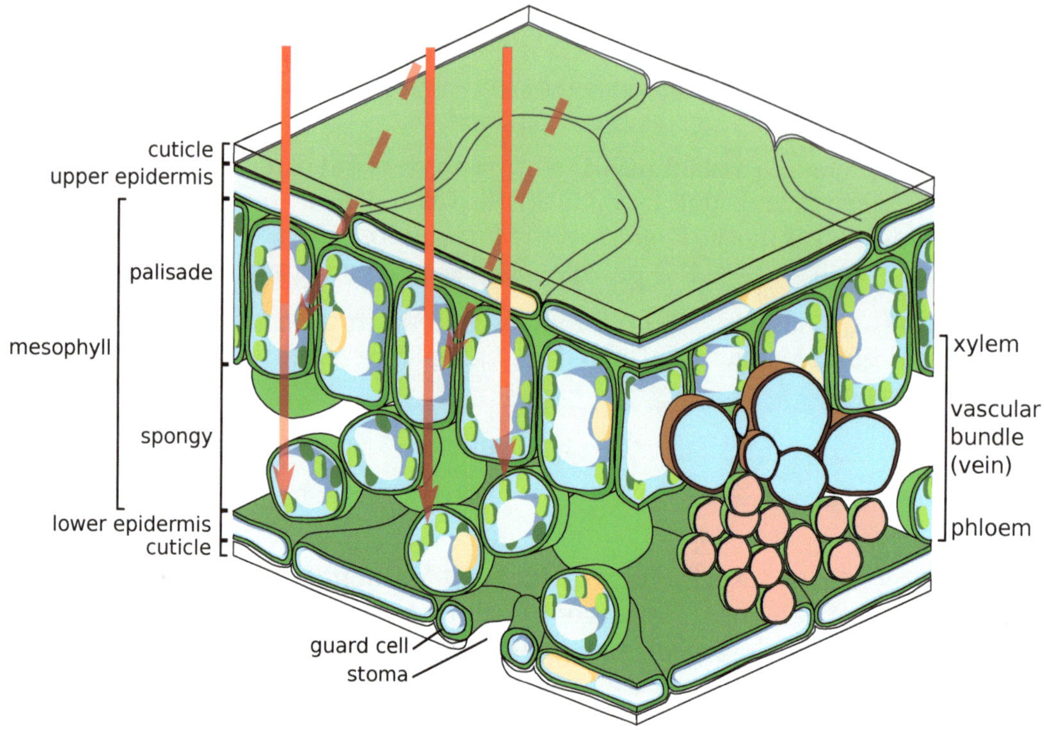

Figure 5.21. Leaf anatomy.

When a typical stem vascular bundle (which has xylem under phloem) enters the leaf, xylem usually faces upwards, whereas phloem faces downwards. Bundles of C_4-plants have additional bundle sheath cells in their vascular bundles.

The epidermis includes typical epidermal cells, stomata surrounded with guard cells (also optionally with subsidiary cells), and trichomes. Almost all epidermal cells are covered with waterproof cuticle, rich of lignin and waxes.

The stomata assists in gas exchange, cooling and water transpiration. There are two guard cells paired together on each side of the stoma. These guard cells are kidney beans shaped and have a thicker cell wall in the middle. The thicker cell wall on the inside makes use of the so-called “bacon effect” (when bacon slice curved on the frying pan) because thinner part of the cell wall is more flexible and therefore bends easier. The same curving effect might be seen in blowing air balloon with the piece

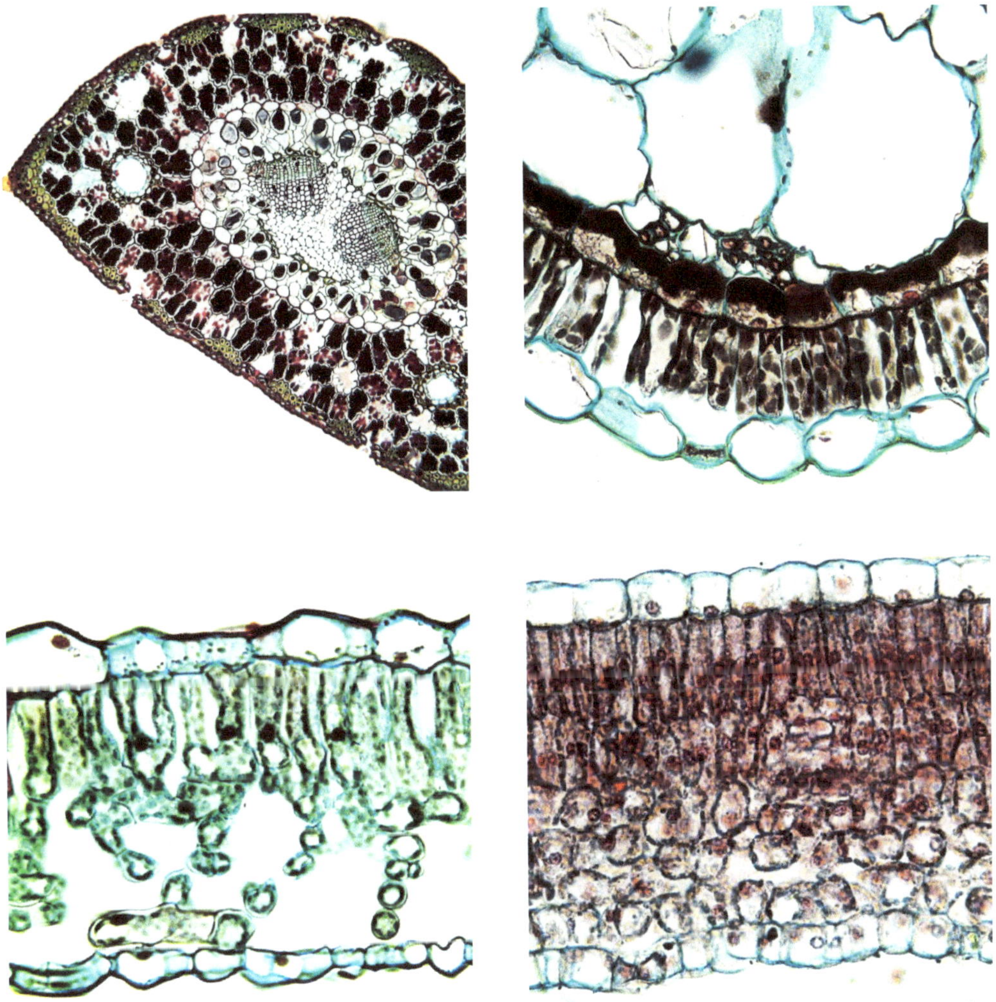

Figure 5.22. Left to right, top to bottom: leaf of sclerophyte *Pinus*, leaf of salt-avoiding (succulent-like) halophyte *Salsola* (epidermis is at the bottom), shade leaf of *Sambucus*, leaf of *Syringa* with guards cells (bottom left). Magnifications $\times 100$ (first) and $\times 400$ (others).

of scotch on one side. The opening of the stoma starts from K^+ accumulation, then osmosis inflates guard cells, and finally the uneven cell wall facilitates the opening of stoma. The stoma closes when the potassium ions exit the cell and water amount decreases in its vacuoles (Fig 5.23).

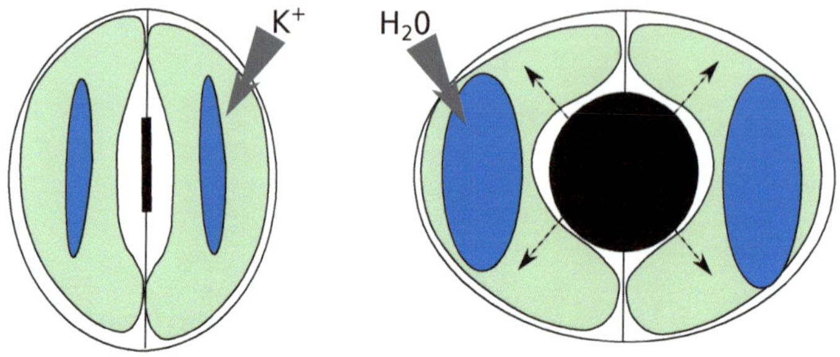

Figure 5.23. Closed and opened stoma. Cell walls are white, cytoplasm green, vacuoles blue.

In most cases, the lower epidermis contains more stomata than the upper epidermis because the bottom of the leaf is cooler and transpiration there is safer. A similar logic is applicable to trichomes (hairs): they are also more frequent on the lower side of the leaf.

5.3.3 Ecological Forms of Plants

When plants adapt to the particular environment conditions, leaves usually respond first. Conversely, one can estimate the ecology of plant simply looking on its leaves.

In regards to water, there are four main types of plants: xerophytes, mesophytes, hygrophytes, and hydrophytes. **Xerophytes** are adapted to the scarce water (Fig. 5.22), they could be **sclerophytes** (usually with prickly and/or rich of sclerenchyma leaves) and **succulents** (with water-accumulating stems or leaves). **Mesophytes** are typical plants which adapt to regular water. **Hygrophytes** live in constantly wet environment, their leaves adapted to high transpiration and sometimes even to guttation (excretion of water drops). **Hydrophytes** grow in water, their leaves are frequently highly dissected to access more gases dissolved in water, and their leaf petioles and stems have air canals to supply underwater organs with gases.

In regards to light, plants could be sciophytes or heliophytes. **Sciophytes** prefer the shade to sunlight, their leaves contain mostly spongy mesophyll. **Heliophytes** prefer the full sun and therefore have leaves filled with palisade mesophyll. The intermediate group are "partial shade" plants.

Halophytes, nitrate halophytes, oxylophytes, and calciphytes are ecological groups adapted to the over-presence of particular chemicals. **Halophyte** plants are frequent, they accumulate (and look similarly to succulents), excrete or avoid (which looks like sclerophyte) sodium chloride (NaCl). They grow in salty places: sea shores, salt deserts and solonets prairies. **Nitrate halophyte** plants grow on soils rich in $NaNO_3$. **Oxylophytes** grow in acidic soils, whereas **calciphytes** grow in basic, chalk soils rich in $CaCO_3$.

Leaves will also reflect adaptations to the substrate, ecological forms named **psammophytes** (grow on sand), **petrophytes** (grow on rocks), and **rheophytes** (grow in fast springs). The latter plants frequently have serious simplifications in their body plan, their leaves and stems are often reduced to form a thallus-like body.

Parasitic plants could be classified in mycoparasites, hemiparasites, and phytoparasites. **Mycoparasitic** plants feed on soil fungi, **phytoparasitic** plants are either plant root parasites or plant stem parasites lacking chlorophyll and photosynthesis. **Hemiparasitic** plants are those which still have chloroplasts but take the significant part of water and even organic compounds from the host plant (like mistletoe, *Viscum*).

5.4 The Stem

The **stem** is an *axial organ of shoot*. It has functions of support, transportation, photosynthesis, and storage. Stem has radial structure, no root hairs and grows continuously.

5.4.1 Morphology of the Stem

Stem morphology is simple. Its components are **nodes** (places where leaves are/were attached) and **internodes**, long or short (in the last case, plant sometimes appears to be stemless, rosette-like).

Stems are different by the type of phyllotaxis. The **phyllotaxis** refers to the arrangement of leaves. If there is one leaf per node, it is a **spiral** (**alternate**) arrangement. Two leaves per node means **opposite** arrangement. Opposite leaves can be all in the same plane or each pair can rotate at 90°. If there are more than two leaves per node, it is a **whorled** arrangement, and each whorl can also rotate.

Each type of spiral phyllotaxis has its own angle of divergence. Multiple types of spiral leaf arrangement mostly follow the **Fibonacci sequence**:

$$\frac{1}{2}, \frac{1}{3}, \frac{2}{5}, \frac{3}{8}, \frac{5}{13}, \frac{8}{21}, \ldots$$

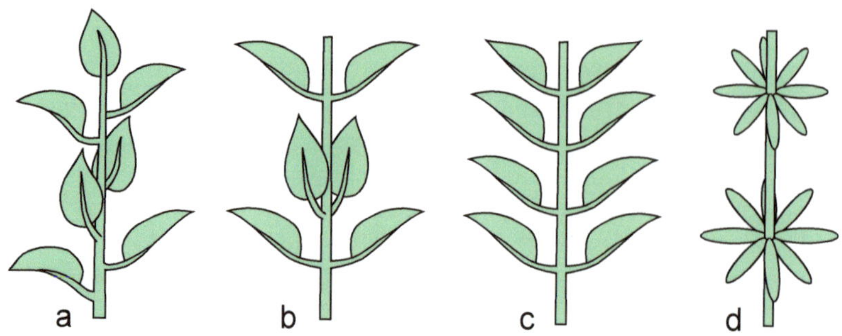

Figure 5.24. Types of phyllotaxis (leaf arrangement): a spiral (alternate), b and c opposite, d whorled.

This sequence of numbers made with simple rule: in the every following fraction, the numerator and denominator are sums of two previous numerators and denominators, respectively. The sequence looks fairly theoretical but amazingly, it is fully applicable to plant science, namely to different types of spiral phyllotaxis (Fig. 5.25).

Figure 5.25. Four first Fibonacci types of spiral phyllotaxis: $\frac{1}{2}$, $\frac{1}{3}$, $\frac{2}{5}$, and $\frac{3}{8}$.

To determine formula of spiral phyllotaxis, one needs to start with arbitrary leaf (or leaf scar) and then find the next (upper) one which is *directed the same way*, lays on the same virtual line. Then, the imaginary spiral should be drawn trough basements from the started leaf to the corresponding upper leaf.

This spiral should go through all intermediary leaves, there might be one, two or more intermediary leaves. Also, the spiral will go at least one time around the stem. (Instead of the imaginary spiral, it is sensible to use a thin thread). One needs to count all leaves in the spiral except the first, and also count number of rotations.

The number of leaves counted will be the denominator of the formula, and the number of rotations is the numerator. This is how Fibonacci numbers appear in plant morphology.

These phyllotaxis formulas are relatively stable and sometimes even taxon-specific. For example, grasses (Gramineae) have $1/2$ phyllotaxis, sedges (*Carex*) $1/3$, many Rosaceae (like apple, *Malus* or cherry, *Prunus*) have $2/5$, willows frequently have $3/8$, *et cetera*.

It is still not absolutely clear why the spiral phyllotaxis is under such a theoretical mathematical rule. The most feasible hypothesis emphasizes mathematical problem of circle packing and the competition between leaf primordia around SAM.

5.4.2 Anatomy of the Primary Stem

Plant evolution resulted first in the **primary stems** with no lateral meristems and secondary tissues. Only long after plants "learned" how to thicken their stems.

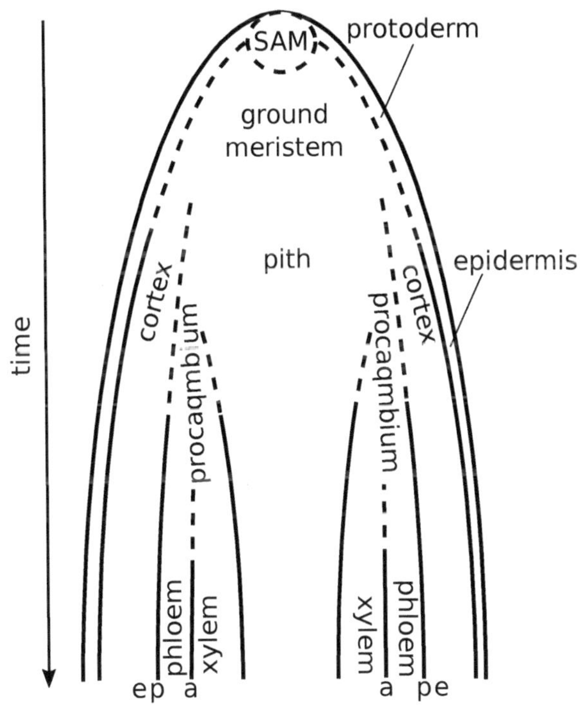

Figure 5.26. Developmental origin of stem tissues (simplified). Letters e, p, a show respectively where **e**ndoderm, **p**ericycle and v**a**scular **ca**mbium might appear.

Development of stem starts from stem apical meristem (SAM) on the top of plant. The SAM produces three **primary meristems**: procambium, protoderm, and ground meristem. **Protoderm** cells differentiate into epidermal cells. The **ground meristem** differentiates into the **cortex** and **pith**. The **procambium** raises between the cortex and the pith. It forms **vascular bundles** or **vascular cylinder** (Fig. 5.26).

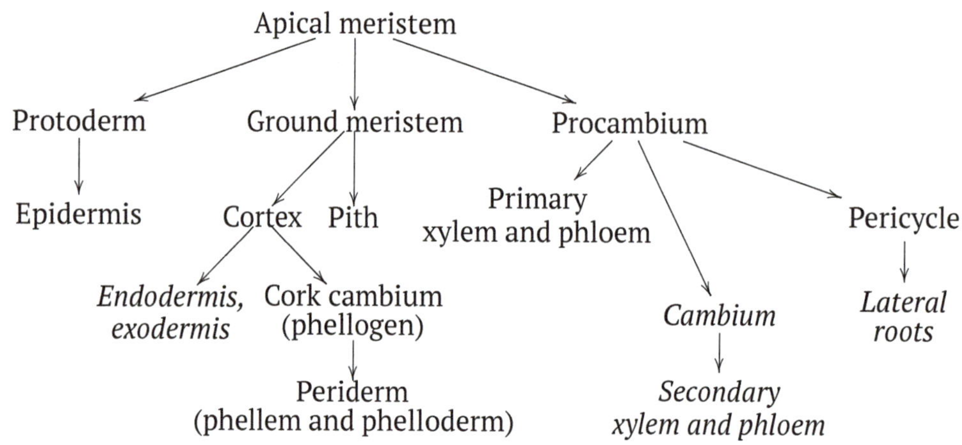

Figure 5.27. Developmental origin of stem tissues (detailed). Root tissues have similar ways of development.

The outer layers of the procambium form the *primary phloem*. The inner layers become the *primary xylem*. The middle layer can be entirely spent or will make cambium for the secondary thickening. At times, the layers of the outside of the procambium can form a **pericycle**. Sometimes the innermost layer of the cortex can form an **endodermis** (endoderm) (Fig. 5.28), and outermost layer makes the **exodermis** (exoderm). All these layers are some kind of the "border control" between functionally different layers of stem. Another frequent variant is the development of collenchyma in the cortex adjacent to epidermis.

Vascular bundles connect leaves and stems. In many plants, they form a ring on the cross-section of the stem. Parenchyma (ground tissue) between vascular bundles typically belongs to both cortex and pith. Another variant is a *vascular cylinder*, structure which fully encircles the stem. Liliid (monocot) stems generally have dispersed vascular bundles. These three variants are **steles**, overall configurations of the primary vascular system of the plant stem (Fig. 5.29). The most frequent kinds of steles are **eustele** (vascular bundles in a ring), **solenostele** (vascular cylinder) and **ataktostele** (dispersed vascular bundles).

All these types were probably originated from **protostele**, configuration where central xylem is surrounded with phloem and no pith is present (Fig. 5.30). While the

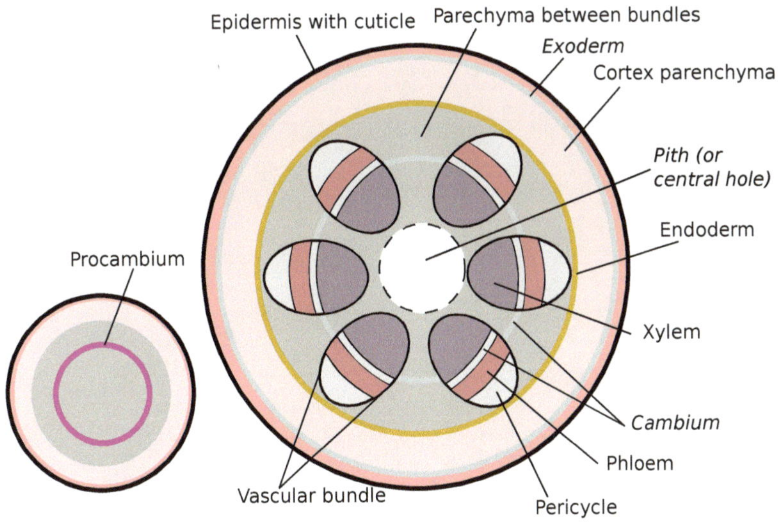

Figure 5.28. Anatomy of the primary stem (right). *Slanted font* is used for "optional" tissues. Small image on the left is the young stem consisted of epidermis, cortex, procambium and pith.

protostele was typical for many prehistoric plants, now only some lycophytes (*Huperzia*) have protostele in stems. Saying that, it is important to note that roots of most plants have vascular tissues arranged similarly to protostele.

5.5 The Root

Root is a latest evolutionary innovation in the vegetative plant anatomy. Many "primitive" plants (all mosses and even some ferns like *Psilotum*) do not have roots; some flowering water plants like the¿ rootless duckweed (*Wolffia*) or the coontail (*Ceratophyllum*) have also reduced their roots. However, large homoiohydric plants need the constant supply of water and minerals, and this evolutionary challenge was responded with appearance of the root system.

Root in *an axial organ of plant with geotropic growth*. One of root functions is to supply anchorage of the plant body in soil or on various surfaces. Other functions include water and mineral absorption and transport, food storage, and communication with other plants.

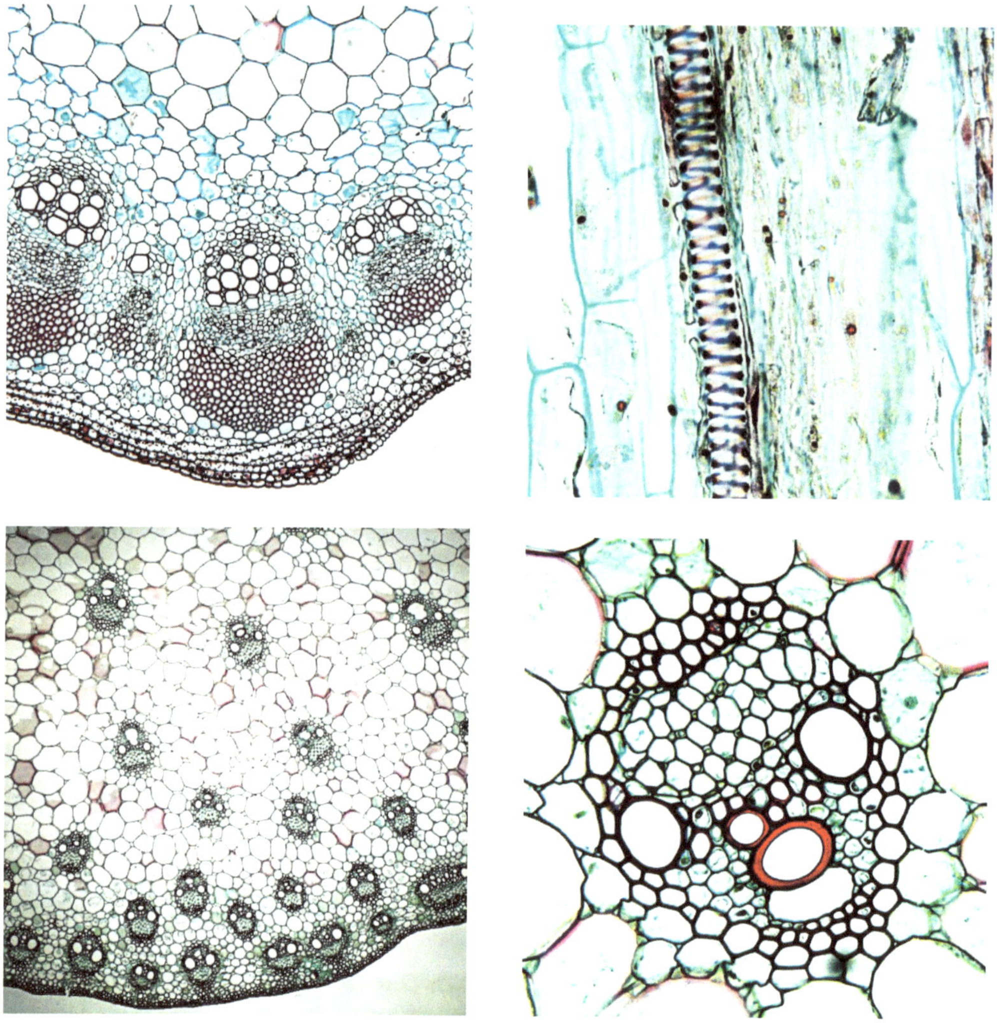

Figure 5.29. Left to right, top to bottom: eustele (stem segment), xylem vessel (longitudinal section) and ataktostele (stem segment and vascular bundle). First photo is from the stem of *Helianthus*, second from *Trifolium*, last two from the stem of *Zea*. Magnifications $\times 100$ (first and third) and $\times 400$ (second and fourth).

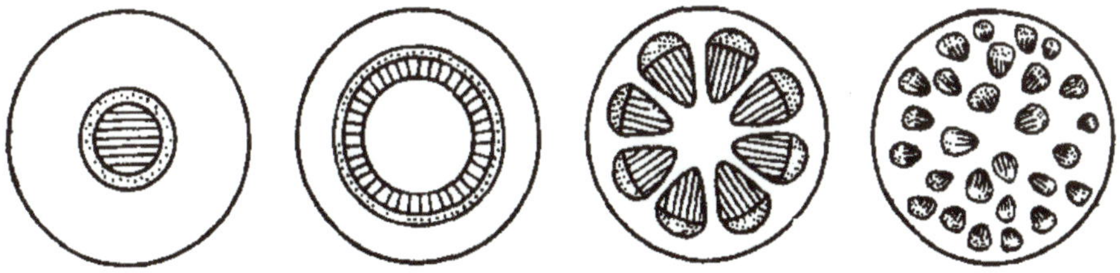

Figure 5.30. Steles (left to right): protostele, solenostele, eustele, ataktostele. Xylem is lined, phloem is dotted.

5.5.1 Morphology of the Root

There are two types of root systems. The first is a **fibrous root system** which has multiple big roots that branch and form a dense mass which does not have a visible primary root ("grass-like"). The other is the **tap root system** which has one main root that has branching into lateral roots ("carrot-like").

Along with having different systems, there are different types of roots: **primary root** originated from the root of the seedling, **secondary (lateral) roots** originate from the primary roots, and **adventitious roots** originate on stems (sometimes also on leaves), the example are prop roots of screw pine (*Pandanus*).

* * *

Roots employ many different modifications which help to protect, interact and storage. For example, roots of parasitic plants are modified into **haustoria** which sink themselves into the vascular tissue of a host plant and live off of the host plant's water and nutrients.

Roots of mangroves (plants growing in ocean coastal swamps) are frequently modified into *supportive aerial roots* ("legs"). Since these swamp plants need oxygen to allow cell respiration in underground parts, there are **pneumatophores**, specialized roots which grow upward (!) and passively catch the air via multiple pores. Plants which grow on sand (psammophytes, see above) have another problem: their substrate constantly disappears. To avoid this, plants developed **contractile roots** which may shorten and pull plant body deeper into the sand.

Some orchid roots are green and photosynthetic (Fig. 5.31)! However, as a rule, root is the heterotrophic organ, because root cells have no access to the light.

Root nodules present on the roots of nitrogen-fixing plants, they contain bacteria capable to deoxidize athmospheric nitrogen into ammonia: $N_2 \rightarrow NH_3$. Root nodules contain also hemoglobin-like proteins which facilitate nitrogen fixation by keep-

Figure 5.31. Photosynthetic roots of leafless orchid *Chiloschista segawai*.

ing oxygen concentration low. Nitrogen-fixing plants are especially frequent among *faboid rosids*: legumes (Leguminosae family) and many other genera (like alder, *Alnus*, or *Shepherdia*, buffaloberry) have root nodules with bacteria. Some other plants (mosquito fern, *Azolla* and dinosaur plant, *Gunnera*) employ cyanobacteria for the same purpose.

Mycorrhiza is a root modification started when fungus penetrates root and makes it more efficient in mineral and water absorption: it will exchange these for organic compounds. In addition to mycorrhizal fungi, **endophytic fungi** inhabit other plant organs and tissues.

5.5.2 Anatomy of the Root

On the *longitudinal section* of young growing root, there are different horizontal layers, zones: root cap covering *division zone*, *elongation zone*, *absorption zone*, and *maturation zone* (Fig. 5.32). The **root cap** protects the root apical meristem (RAM), which is a group of small regularly shaped cells. A small, centrally located part of the RAM is the **quiescent center** where initial cells divide and produce all other cells of root. Root cap is responsible for the geotropic growth, if the root tip comes into contact with a barrier, root cap will feel it and will grow on a different direction to go around it.

The **elongation zone** is where the cells start to elongate, giving it length. The **absorption zone** is where the rhizodermis tissue (root hairs) develops and where water and nutrients are absorbed and brought into the plant. Within the **maturation zone**,

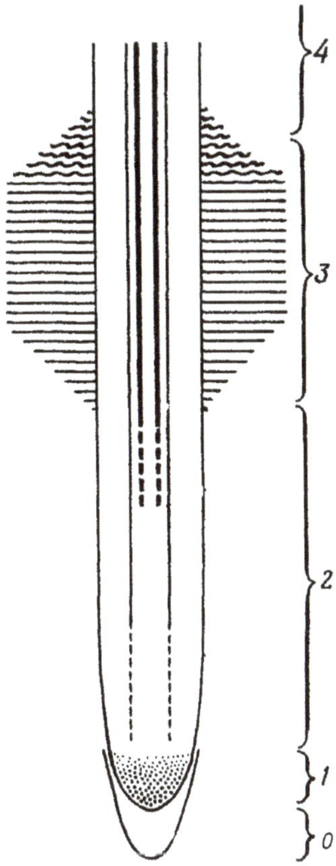

Figure 5.32. Root zones: 0 root cap, 1 division zone, 2 elongation zone, 3 absorption zone, 4 maturation zone.

root hairs degrade, many cells start to acquire secondary walls and lateral roots develop (Fig. 5.32).

On the *cross-section* of the root made within absorption zone, the first tissue is the rhizodermis, which is also known as the root epidermis, then cortex, which segregates external *exodermis* and internal *endodermis* one-cellular layers, and vascular cylinder (Fig. 5.33). Typically, roots have no pith. In some cases (for example, in orchids), cortex may give multi-layered *velamen* (see above), another absorption tissue.

Vascular cylinder is located in the center of the root, it contains the pericycle which is made of mostly parenchyma and bordering endodermis. Pericycle cells may be used for storage, they contribute to the vascular cambium, and initiate the development of lateral roots. Consequently, lateral roots are developing endogenously and break

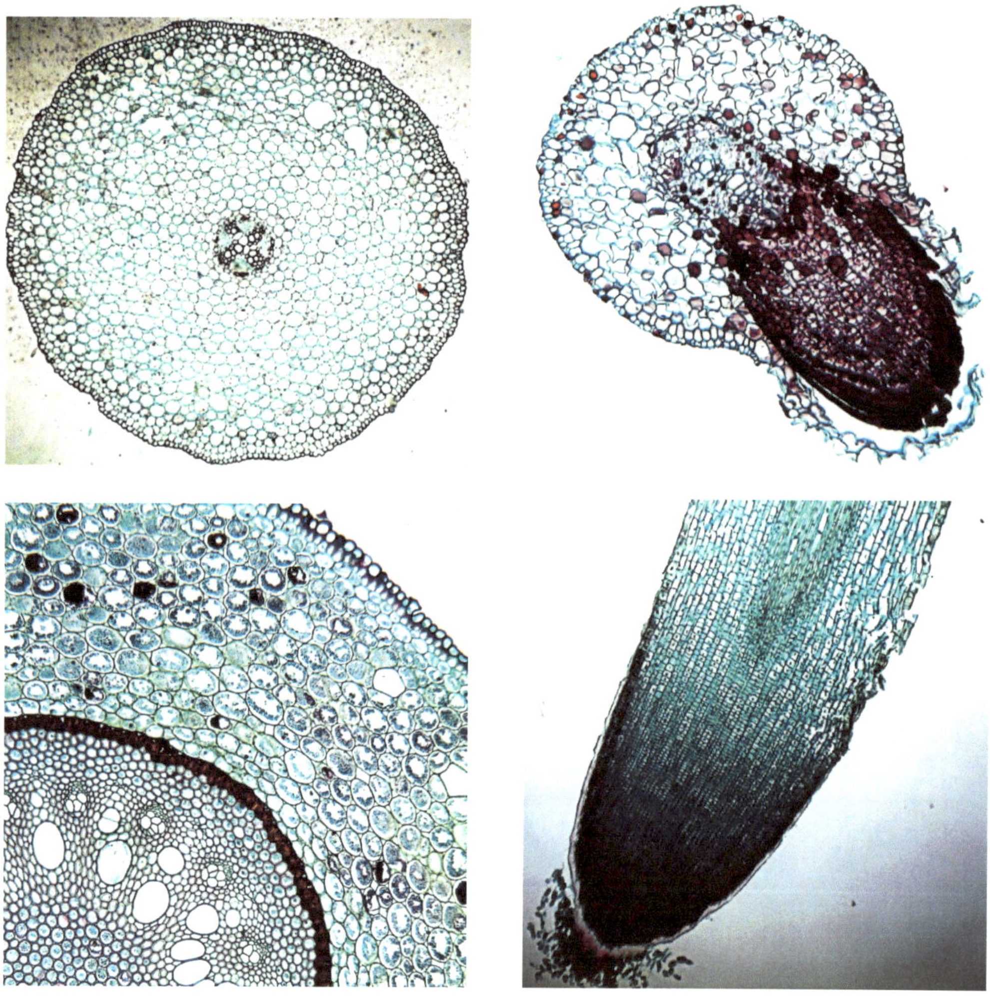

Figure 5.33. Left to right, top to bottom: *Ranunculus* root with 4-rayed xylem, *Salix* root with the lateral root developing, *Smilax* root with visible Casparian stripes in the endodermis; *Zea* root longitudinal section with root cap, division and elongation zones. Magnifications $\times 100$ (second) and $\times 40$ (others).

tissues located outside, like aliens in the famous movie. Root phloem is arranged in several strands whereas xylem typically has a radial, sometimes star-shaped structure with few rays (Fig. 5.34). In the last case, phloem strands are located between rays of xylem.

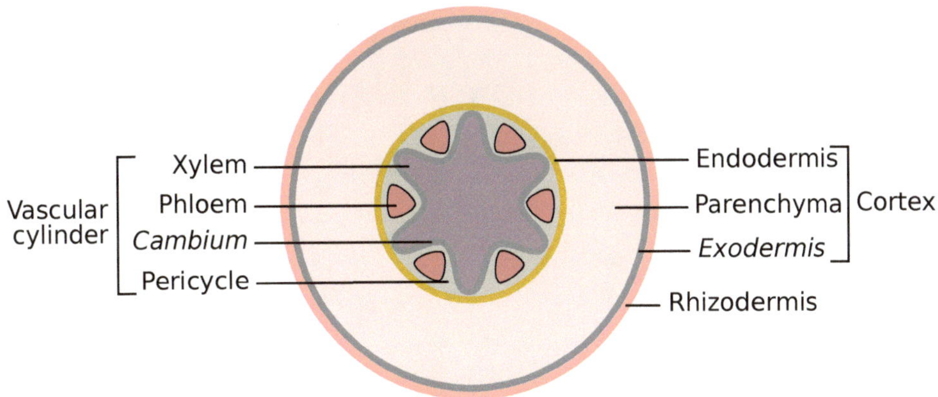

Figure 5.34. Anatomy of root: cross-section through the maturation zone.

Root tissues develop in the way similar to stem, RAM gave rise to ground meristem, procambium, and the protoderm, which in turn make all primary tissues mentioned above. Later, pericycle develops into lateral roots or the vascular cambium which in turn produces into the secondary xylem and phloem. The secondary root is similar to secondary stem (see below).

5.5.3 Water and Sugar Transportation in Plants

Plants need water to supply photosynthesis (the oxygen is from water!), to cool down via transpiration, and to utilize diluted microelements. Dead velamen (paper-like), rhizoids (hair-like), and living rhizodermis (rhizoderm) are responsible for water uptake.

In rhizodermis, root hairs increase the surface area where the plant has to absorb the nutrients and water. To take water, hair cells increase concentration of organic chemicals (the process which needs ATP) and then use osmosis. There are two ways that water transport may go: apoplastic or symplastic. *Apoplastic* transport moves water through the cell walls of cortex: from the rhizodermis to the endodermis. Endodermis cell walls bear **Casparian strips** (rich of hydrophobic suberin and lignin) which prevent the water from passing through the cell wall and force *symplastic* transport (Fig 5.35) through cytoplasms and plasmodesmata. Symplastic transport there is directed to the center of root only and requires ATP to be spend.

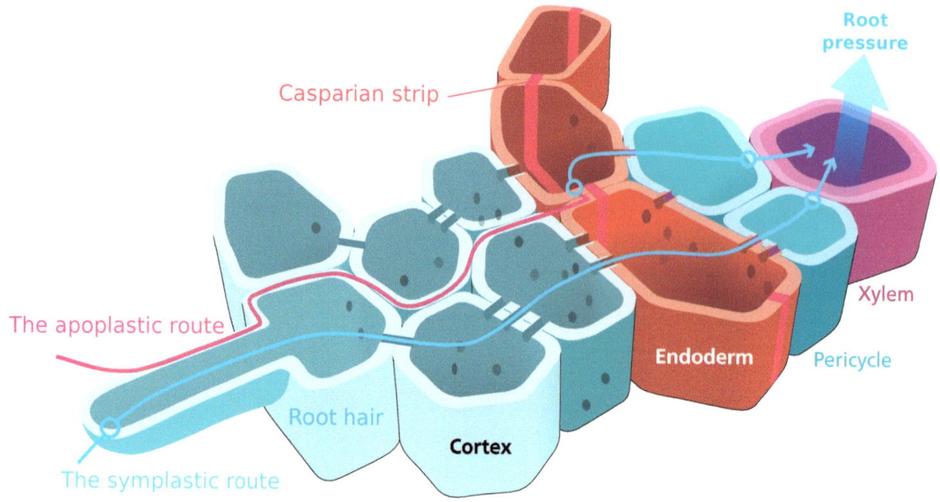

Figure 5.35. Symplastic and apoplastic transport in root.

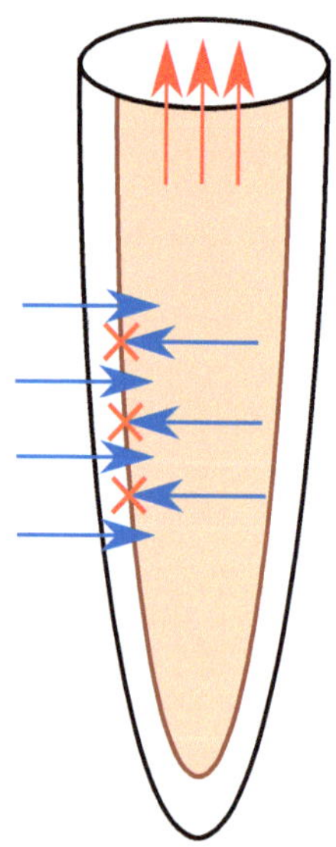

Figure 5.36. The origin of root pressure: water comes into vascular cylinder but cannot go back because of endoderm (brown line). The only possible way is to go up.

By pumping water inside vascular cylinder and not letting it back, endodermis cells create the **root pressure** (Fig. 5.36). It is easy to observe on tall herbaceous plants cut near the ground: drops of water will immediately appear on the cutting. Inside tracheary elements of xylem, water moves with the root pressure, capillary force and the sucking pressure of transpiration. The latter means that water column does not want to break and if water disappears from the top (stomata on leaves), it will move water inside plant. The main direction of water movement is from roots to leaves, i.e. upwards.

Products of photosynthesis (sugars) are moving inside living cells of phloem; these cells (sieve tubes) use only symplastic transport to distribute glucose and other organic compounds among all organs of plants. In fact, phloem transports these components in all directions: to the flowers (usually upwards), and at the same time to the roots (usually downwards).

Chapter 6

Growing Diversity of Plants

When plants developed basic tissues and organs and thus became mature enough to survive on land, they started to increase in their diversity. All plants studied in this and following chapters belong to plants$_2$, or kingdom Vegetabilia which is split into three phyla (Fig. 6.1): Bryophyta (mosses and relatives), Pteridophyta (ferns and allies), and Spermatophyta (seed plants). The most striking differences between these phyla lay in the organization of their life cycles.

Land plants have a sporic life-cycle (Fig. 4.13) that begins with a diplont (sporophyte); the mother cell of spores goes through the meiosis and produces haploid spores. These spores develop into haplont which produces female and male gametangia (gamete "homes"). Female is called *archegonium*, the male—*antheridium*; the archegonium produces oocyte which is fertilized by the antheridium's spermatozoon in the process of oogamy. When this fertilization happens, it forms a diploid zygote which then matures into a *young sporophyte growing on a gametophyte*. This kind of same species parasitism is almost unique in the living world. Only viviparous animals (like mammals with their pregnancy) could be compared with land plants.

6.1 Bryophyta: the mosses

Bryophyta has gametophyte predominance while Pteridophyta and Spermatophyta both have sporophyte predominance (and the main difference between Pteridophyta and Spermatophyta is that Spermatophyta has seeds). Bryophyta has approximately 20,000 species. They do not have roots, but have long dead cells capable of water absorbency via apoplastic transport, these cells are called **rhizoid cells**. Their sporophyte is reduced to **sporogon**, which is simply a sporangium with **seta** (stalk), and is usually parasitic. Gametophyte of bryophytes starts its development from a **pro-**

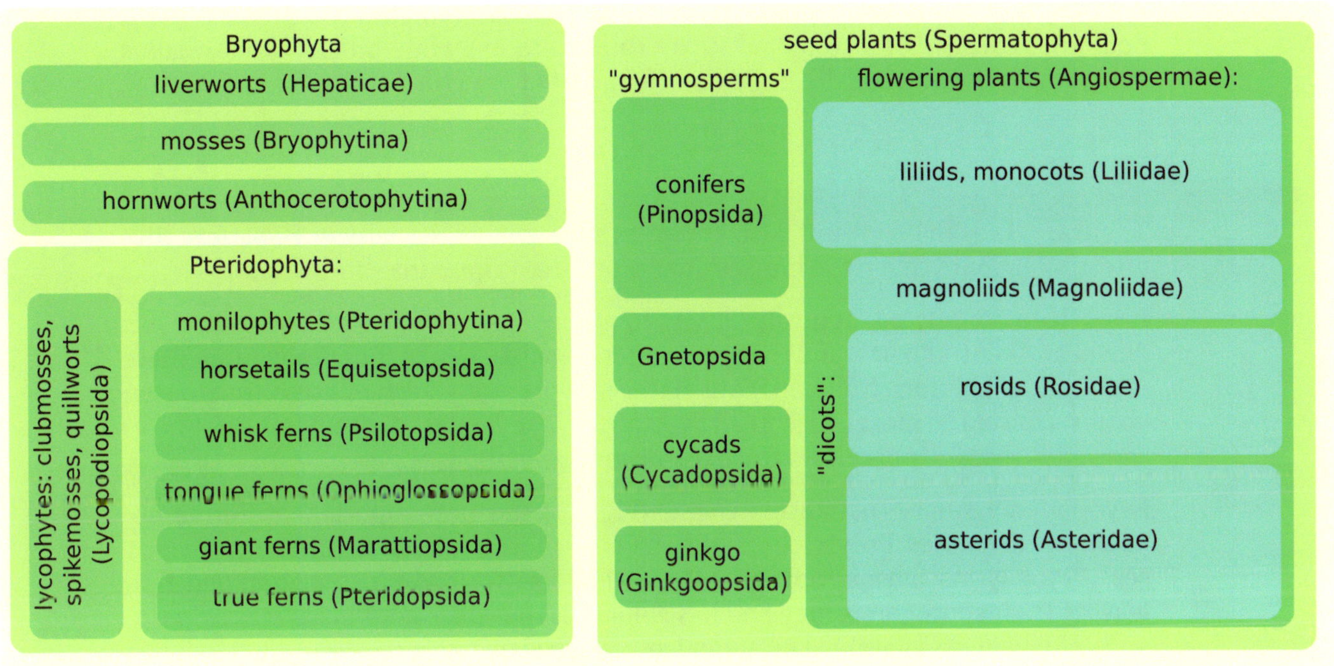

Figure 6.1. $Plants_2$ classification: detailed scheme.

tonema, thread of cells. Bryophyta are poikilohydric; they go through dehydration or extremely low water concentration without any serious physiological damage to the plant.

Life cycle of mosses is similar to the general life cycle of land plants described above. They begin with a gametophyte with an archegonia and antheridia. The antheridium produces biflagellate spermatozoa which fertilizes the egg and produces diploid zygote; zygote grows into a sporogon and its cells (mother cells of spores) go through meiosis which produces haploid spores. Spores will be distributed with the wind, land on the substrate and germinate into protonema stage which then develops into a green, well-developed gametophyte. Most of moss gametophytes have a shoot body that consists of a stem and leaves (but no roots) while others have a thallus body, which is a flat, leaf-like, and undifferentiated structure.

* * *

There are three main groups, also known as subphyla, of Bryophyta: Hepaticae (liverworts), Bryophytina (true mosses), and Anthocerotophytina (hornworts).

Hepaticae are phylogenetically closest to green algae. Their thallus typically has dorsal and ventral parts, and the sporogon is bag-like. Inside the sporangium, there is no central column (columella) but elaters are present, which are cells that loosen spores. One of the most widespread liverwort is *Marchantia*, it is commonly found in wet shady places. It became a frequent weed in greenhouses.

Bryophytina consists of multiple classes (Fig. 6.2), the most important are **Sphagnopsida**—peat mosses, **Polytrichopsida**—hair cap mosses, and **Bryopsida**—green mosses. Bryophytina have a radially structured shoot-like body with a stem and thin leaves. Their sporogon is long and has columella, but does not have elaters. Sporogons of true mosses are usually supplied with **peristome**, structure which helps in spore distribution. Some advanced true mosses (hair cap moss, *Polytrichum*) have tall gametophyte with proto-vascular tissues, while others (stinkmoss, *Splachnum*) employ insects for the distribution of spores. Peat moss (*Sphagnum*) is probably the most economically important genus of Bryophyta.

Anthocerotophytina (Fig. 6.3) evolutionary are closests to the next phylum, Pteridophyta (ferns and allies). Hornworts have a flattened thallus body, their long photosynthetic sporogon has columella and elaters. The presence of stomata on sporogons and the ability of some hornwort sporogons to branch and sometimes even live independently from the gametophyte provide a support for the advanced position of this group. Hornworts are rare and quite small (first millimeters in size), and like liverworts, they prefer shady and wet places.

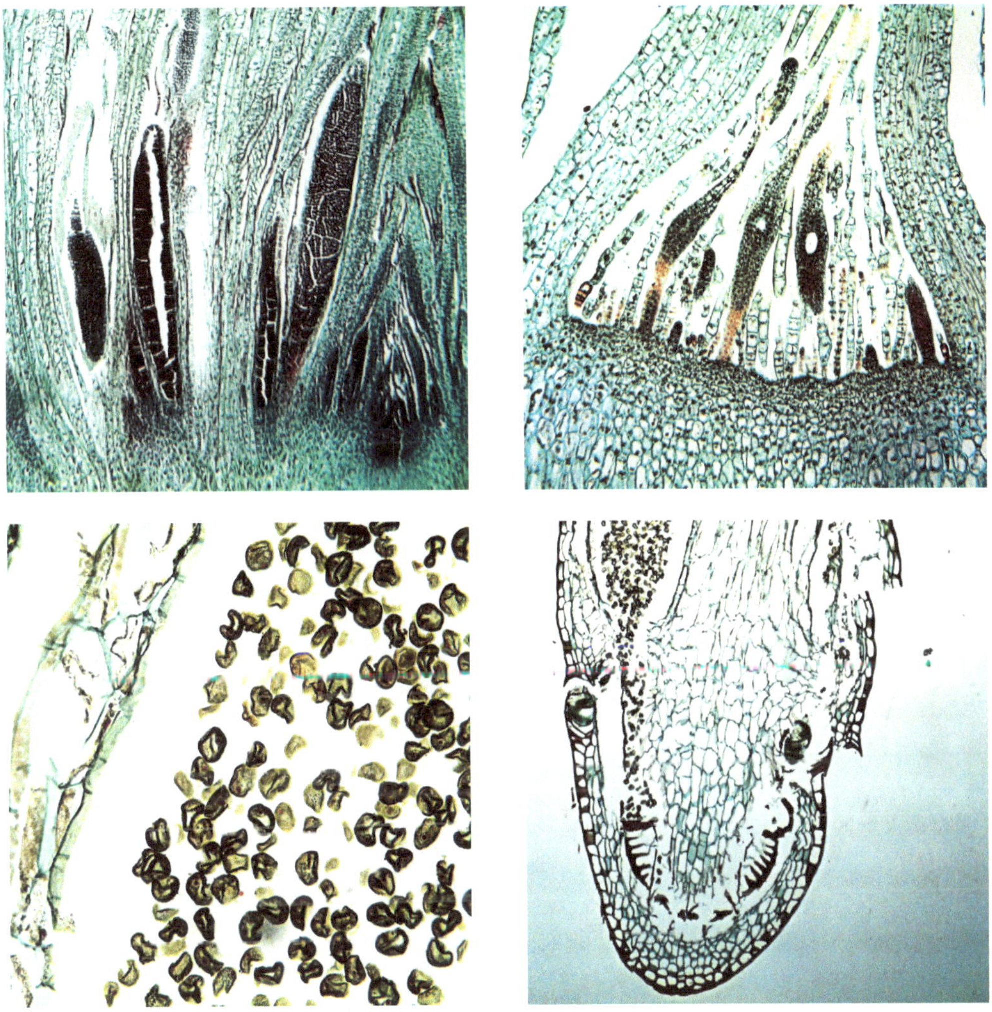

Figure 6.2. Left to right, top to bottom: *Mnium* (Bryopsida) antheridia, archegonia, spores and the base of sporogon. Magnifications ×100 (first and second) and ×400 (third) and ×50 (fourth).

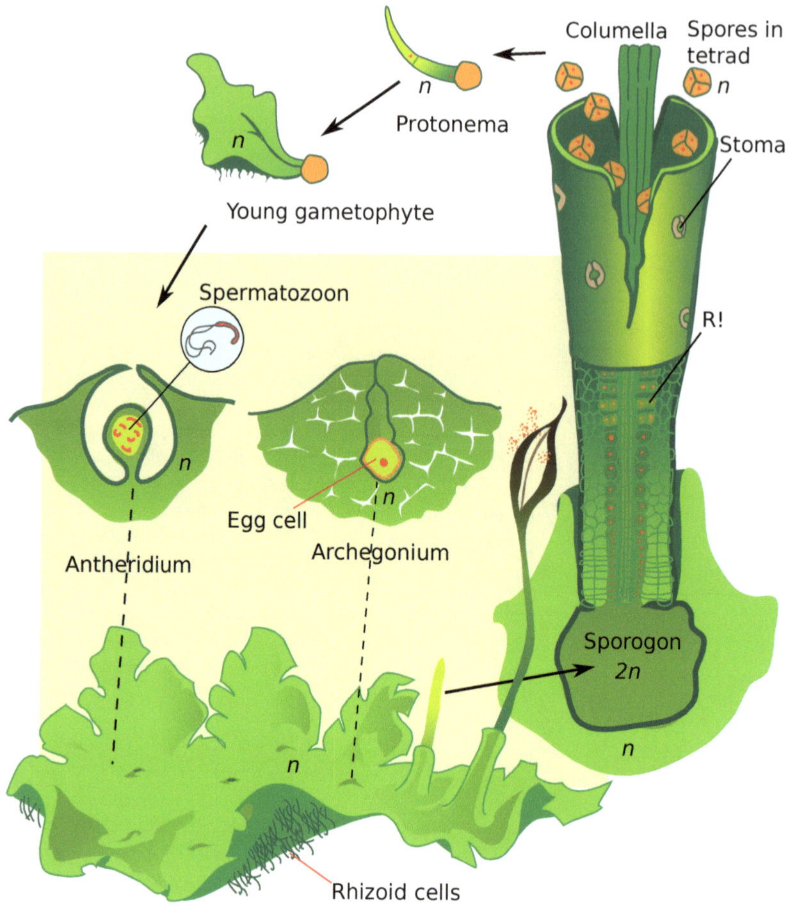

Figure 6.3. Life cycle of *Phaeoceros* (Anthocerotophytina).

Mosses have become known as the "evolutionary dead end" because their poikilohydric gametophyte requires water for fertilization and does not have a root system; this restricts the size and requires dense growing. However, if the sexual organs are near the soil surface, then the parasitic sporogon would not grow tall enough, and consequently would not be able to effectively distribute spores with the wind.

Three natural forces "tear" the body of moss: wind and light require plant to be taller whereas water requires it to be smaller (Fig. 6.4). Mosses did not resolve this conflict.

The only way to fix the situation properly would be to make the sporophyte taller, independently growing and therefore and reduce dominance of the gametophyte. This is what ferns (Fig. 6.5) did.

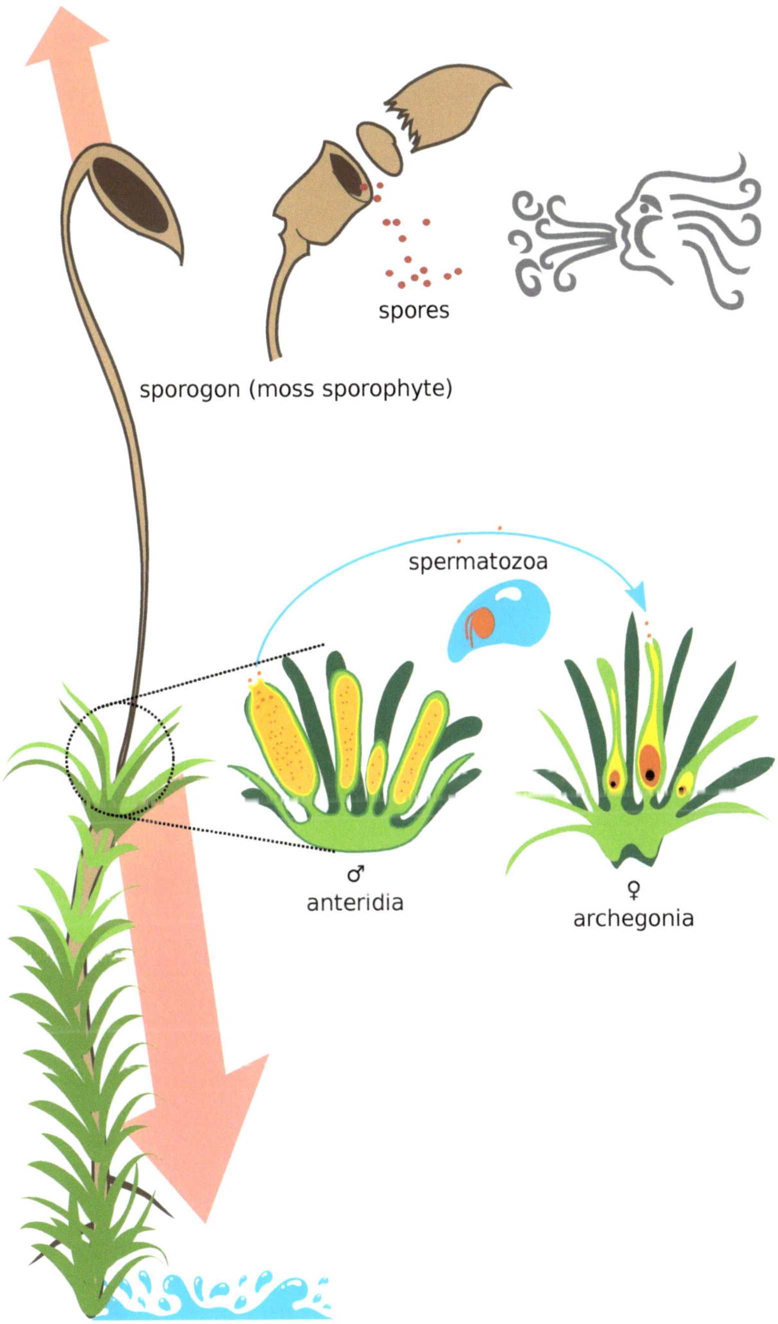

Figure 6.4. Two forces which disrupt moss evolution.

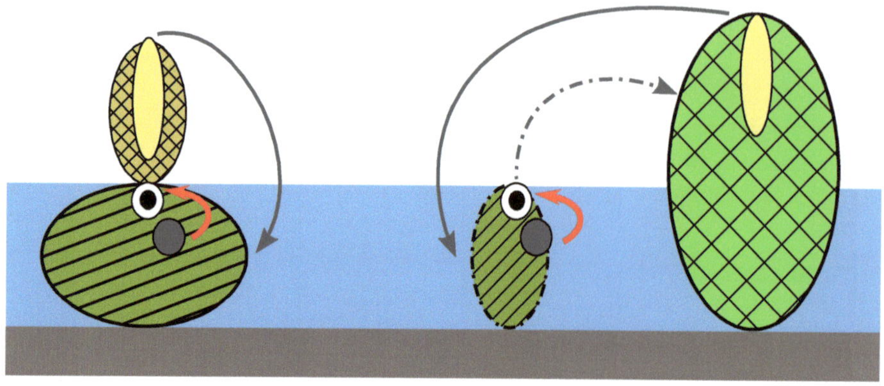

Figure 6.5. Mosses (left) *vs.* ferns (right). Water level is blue, sporangia are yellow, gametangia gray. Red arrows show fertilization (in water), grey arrows designate life cycles. Dotted lines is *independence* of fern sporophytes from their ephemeral gametophytes.

6.2 Pteridophyta: the ferns

Pteridophyta, ferns and allies, have approximately 12,000 species and six classes (Fig.6.6). They have a sporic life cycle with sporophyte predominance whereas their gametophytes are often reduced to prothallium, small hornwort-like plant. Another frequent variant is the underground, mycoparasitic gametophyte. Pteridophyta (with one exception) have true roots. Most of them have vascular tissues and are homoiohydric. This is why seed plants together with ferns have a name **vascular plants**. Pteridophyta sporophytes always start their life from an embryo located on the gametophyte. While Pteridophyta have true xylem and phloem, they do not have developed secondary thickening.

Most ancient pteridophytes appeared in Silurian period, they were rhyniophytes. Rhyniophyles had well-developed aboveground gametophytes and relatively short, dichotomously branched leafless sporophytes. The next important steps were formation of leaves and further reduction of gametophytes.

6.2.1 Diversity of pteridophytes

Lycopodiopsida, or lycophytes have at least four genera and more than 1,200 species. Lycophytes belong to **microphyllous** lineage of pteridohytes. This means that their leaves originated from the emergences of the stem surface, and therefore are more similar to moss leaves than any other leaves of pteridophytes and seed plants. Lycophyte sporangia are associated with leaves and often form **strobilus** which is a condensation of sporangia-bearing leaves (**sporophylls** when they are leaf-like or **sporangiophores** when they are divergent). Their spermatozoon usually has 2 flag-

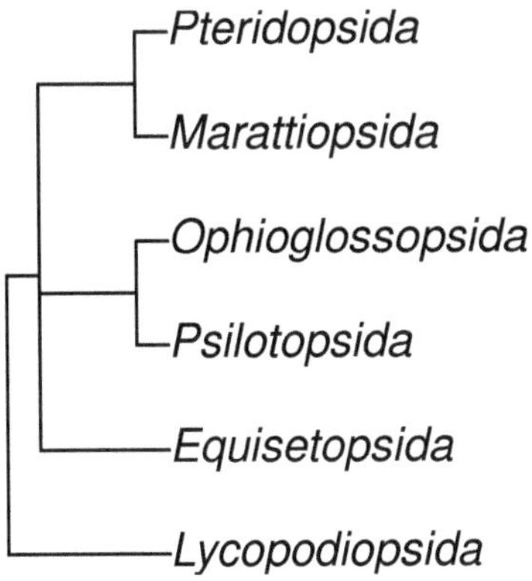

Figure 6.6. Phylogeny of the six classes of Preridophyta.

ella (like mosses) but are sometimes also multiflagellate (like spermatozoa of other ferns). Lycophytes used to be the dominant plants of Carboniferous tropical swamp forests and their remains became coal. Contemporary lycophytes are much smaller but still thrive in wet and warm places. More basal lycophytes (clubmosses *Huperzia* and *Lycopodium*) have equal spores and underground gametophytes, whereas more advanced *Selaginella* (spikemoss) and *Isoëtes* (quillwort) are both heterosporous (see below) with reduced aboveground gametophytes. Quillwort is a direct descendant of giant Carboniferous lycophyte trees, and despite being an underwater hydrophyte, it still retains the unusual secondary thickening of stem. Many spike mosses are poikilohydric (another similarity with mosses).

Equisetopsida (horsetails) is a small group with one genus, *Equisetum*, and has about 30 different herbaceous species that typically live in moist habitats. The leaves of these plants are reduced into scales, and the stems are segmented and also photosynthetic; there is also an underground rhizome. The stem epidermis contains silica which makes it have an abrasive surface, and because of this, American pioneers would use this plant to scour pots and pans. This is how it received the nickname "scouring rush." The stem has multiple canals, this is somehow similar to stems of grasses. The sporangia are associated with hexangular stalked sporangiophores; there are also elaters which are not separate cells but parts of the spore wall. Gametophytes are typically minute and dioecious, but the plants themselves are homosporous: smaller suppressed gametophytes develop only antheridia while larger gametophytes develop only archegonia.

Psilotopsida (whisk ferns) is a small tropical group which consists of only two genera, *Psilotum* and *Tmesipteris*, with only seven different species. They are herbaceous plants that grow as epiphytes. Whisk ferns are homosporous, and their sporangia are fused into **synangia**. Psilotopsida have protostele like the some lycophytes, and long-lived underground gametophytes; they also have multiflagellate spermatozoa similar to all other ferns. Both *Psilotum* and *Tmesipteris* lack roots; in addition, *Psilotum* also lacks leaves.

Ophioglossopsida (tongue ferns) is a small group that consists of approximately 75 species, and are closest relatives to whisk ferns. Ophioglossopsida have an underground rhizome (sometimes with traces of secondary thickening) with aboveground bisected leaves: one half of each leaf is the leaf blade while the other half becomes the *sporophyll*. The gametophytes also grow underground. *Ophioglossum vulgatum*, known also as the adder's tongue fern, has chromosome number $2n$ =1,360 which is the largest chromosome number ever!

Marattiopsida (giant ferns) are tropical plants, with several genera and about 100 species. These are similar to true ferns and have compound leaves that are coiled when young. They are also the biggest ferns, as one leaf can be six meters in length. They have short stems, and leaves with stipules. Their sporangia have multi-layer walls and are fused into synangia (not like true ferns). At the same time, they are located on the bottom surface of leaves (like in true ferns). Gametophytes are relatively large (1–2 cm), photosynthetic, and typically live for a long time. These ferns were important in the Carboniferous swamp forests.

Pteridopsida (true ferns) have more than 10,000 species and make up the majority of living **monilophytes** (all classes of Pteridophyta except lycophytes). Their leaves are called **fronds** because of apical growth; young leaves are coiled into **fiddleheads** (Fig. 6.7). True ferns are **megaphyllous**: their leaves originated from flattened branches. True ferns have unique sporangia: **leptosporangia**. Leptosporangia originate from a single cell in a leaf, they have long, thin stalks, and the wall of one cell layer; they also open actively: when sporangium ripens (dries), the row of cells with thickened walls (**annulus**) will shrink slower than surrounding cells and finally would break and release all spores at once. Leptosporangia are also grouped in clusters called **sori** which are often covered with umbrella- or pocket-like **indusia**. Gametophytes of Pteridopsida are minute and grow aboveground. Some genera of true ferns (like mosquito fern *Azolla*, water shamrock *Marsilea* and several others) are heterosporous.

True ferns are highly competitive even to angiosperms. In spite of their "primitive" life cycle, they have multiple advantages: abilities to photosynthesize in deep shade (they are not obliged to grow fast), to survive high humidity, and to make billions of reproductive units (spores). Ferns do not need to spend their resources on flowers and fruits, and are also less vulnerable to vertebrate herbivores and insect pests, probably because they do not employ them as pollinators and, therefore, can poison tissues against all animals.

Figure 6.7. Selected stages of *Cystopteris* life cycle, representative of Pteridopsida.

6.2.2 Heterospory: Next step on land

Vertebrate animals became fully terrestrial (amphibians became first reptiles) only when their fertilization became completely independent from water. Plants started to perform the similar "evolutionary efforts" even earlier, but while reptiles actively approach the sexual partner, plants cannot do the same because their tissues and organs evolved for completely different purposes. Instead of the active sex, plants use "carpet bombing" with spores; this was invented to increase the chance that two spores land nearby and the distance between sperm and egg cell will be minimal.

However, since simple increase in the number of spores is a great waste of resources, plants minimized spore size; this will also allow for the longer distance of dispersal. On the other hand, some spores must remain large because embryo (if fertilization occurs) will need the support from the feeding gametophyte. Consequently, plants ended up with division of labor: numerous, minuscule male spores which grow into male gametophytes with antheridia only, and few large female spores which make female gametophytes producing only archegonia(Fig. 6.8).

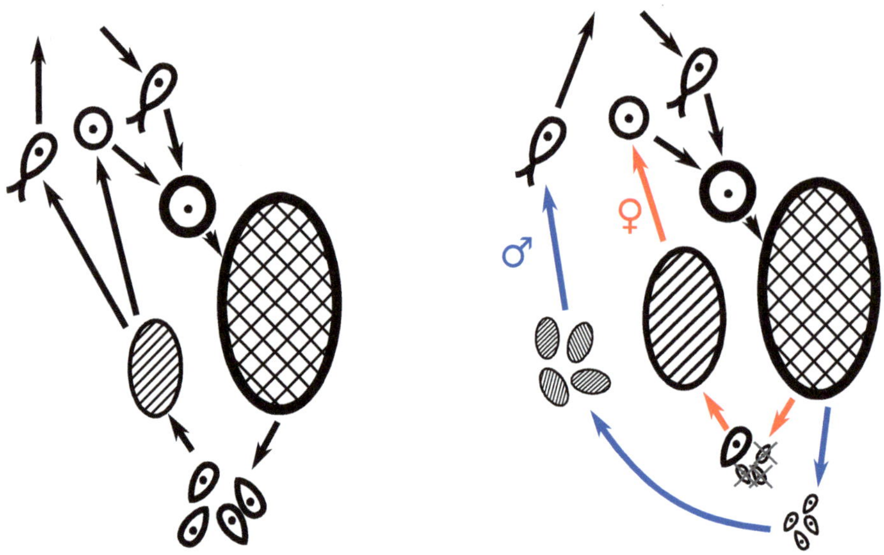

Figure 6.8. From homosporous to heterosporous life cycle.

This **heterosporic** cycle makes fertilization less dependent on water and more dependent on spore distribution and gametophyte features (Fig. 6.9). It also allows for numerous improvements in future.

Division of labor allows resources to be used more efficiently and also restricts self-fertilization. In the plant evolution, there was a high need for heterospory because it independently arose in several groups of pteridophytes and even among mosses. In the extreme cases of heterospory (Fig. 6.10), a female spore does not leave the mother plant and germinate there, "waiting" for the fertilization from the male gametophyte developed nearby; in fact, this is incipient pollination, the step towards the *seed*.

Heterosporous plants produce one female spore, **megaspore**, which is rich in nutrients; megaspores are not widely dispersed, but the female gametophyte that comes of it provides nutrition and protection for the zygote, embryo, and young sporophyte. Heterosporic life cycle (Fig. 6.11) starts with a male gametophyte and a female gametophyte, both of which produce gametes. Once fertilization occurs, a zygote develops into sporophyte. The sporophyte will then produce two different sporangia types: female **megasporangia** and male **microsporgangia**. Meiosis in megasporangium will frequently result in *one* female spore, megaspore (similar to the meiosis in the ovaries of vertebrate animals), whereas in the microsporangium, meiosis

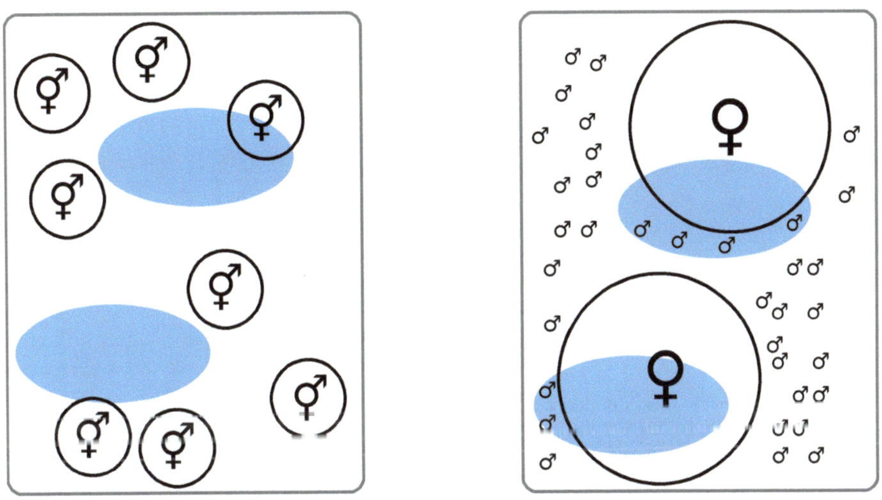

Figure 6.9. Simple scheme which illustrates the heterosporic way of fertilization. Two drops of water (blue) do not provide the connection between two gametophytes of homosporous plant (left) but are enough for gametophytes of heterosporous plant (right) using the same amount of resources.

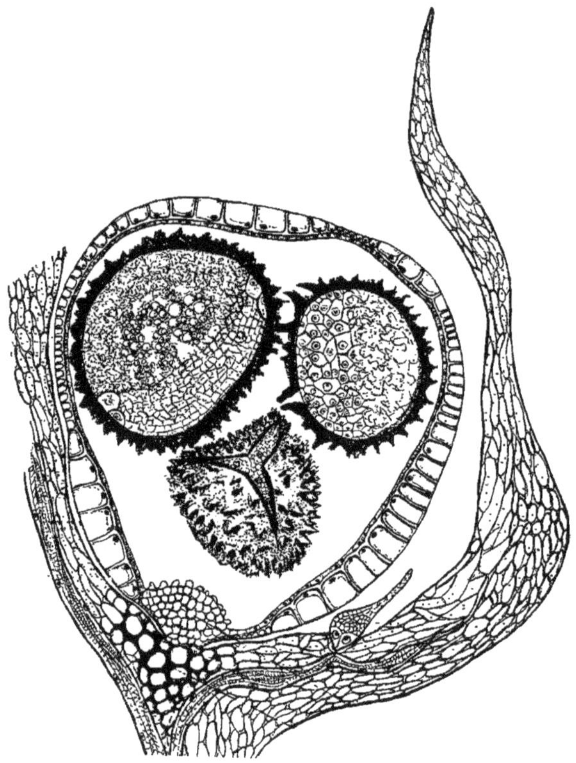

Figure 6.10. Megasporangium of the meadow spike-moss, *Selaginella apoda (from Lyon, 1901)*. All three megaspores germinate into female gametophytes without leaving sporangium.

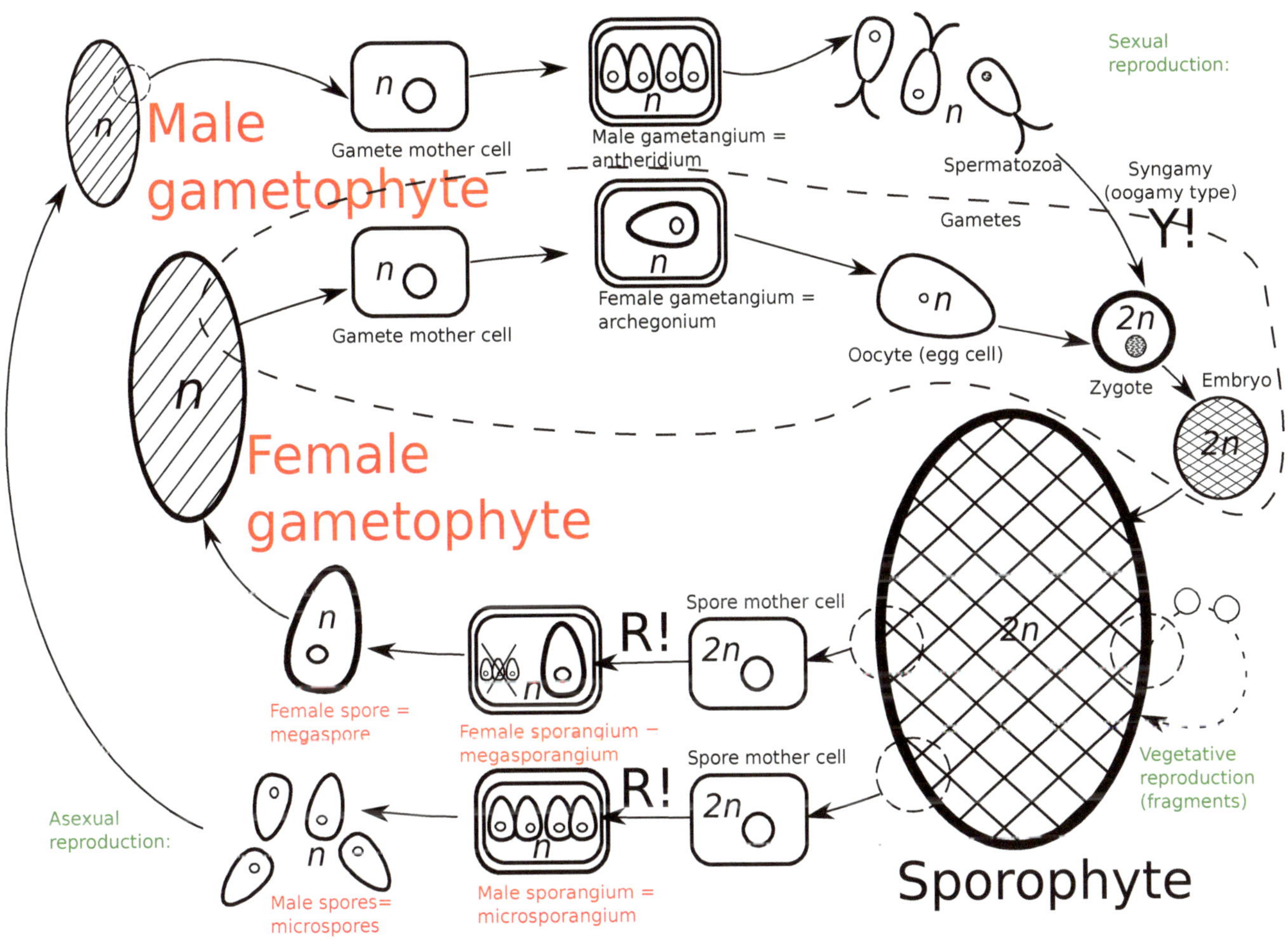

Figure 6.11. Life cycle of heterosporic plant. Innovations (comparing with the life cycle of land plants) are in red.

and subsequent mitoses will make numerous **microspores**; both the megaspore and microspores will develop into gametophytes and the cycle will repeat.

In all, heterospory allows for separation between male and female haploid lineages. Male gametophytes become so small that they could easily be transported as a whole. Whole male gametophytes start to be a moving stage—this is origin of *pollination*.

Chapter 7

The Origin of Trees and Seeds

Competition over resources (primarily water and sun light) always drove plant evolution. The most logical way to escape competition was to enlarge the body. But if only primary tissues are available, this growth is strictly limited.

Without secondary thickening, the trunk will easily break under the weight of growing crown, and the plant will die. This is easy to see in plants which still dare to develop the tree-like habit without secondary growth: tree ferns and palms. In addition, tree ferns have no bark which limits their distribution to the really wet places.

On the other hand, thickening of stem will allow for branching, and branching allow for even bigger aboveground body. But then, new problems associated with both size and life cycle will pose another great challenge.

7.1 Secondary Stem

In many seed plants, secondary growth begins in their first year within the stem and continues on for many more years. These plants are classified as *woody*. They develop *secondary tissues* like periderm and wood, and even *tertiary* structures like bark.

The first step in producing secondary phloem and xylem (other names are metaphloem and metaxylem) is to form the vascular cambium, which involves cell division inside the vascular bundles and the parenchyma that are between the bundles (Fig. 7.1). The vascular cambium divides in two directions. The cells that are formed to the outside become the *secondary phloem*, and those formed to the inside are the *secondary xylem* (Fig. 7.3). After several years, central pith disappears under the pressure of growing wood, and only traces of primary xylem (protoxymem) can be seen *under* the thick

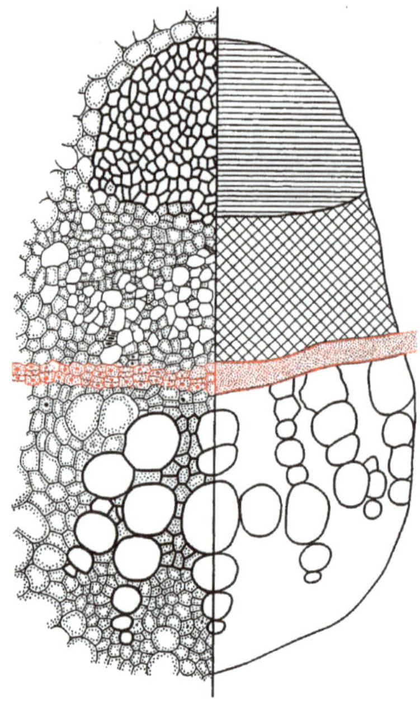

Figure 7.1. Vascular bundle on the stage of cambium (red) formation. Xylem is located downward, phloem upward. Note that cambium forms also *between* vascular bundles.

secondary xylem. Altogether, these tissues (pith + primary xylem + secondary xylem) are **wood** (Fig. 7.2).

The secondary phloem forms outside of the vascular cambium, and traces of primary phloem (protophloem) are visible above it. It is rich in fibers, and unlike the wood, it does not form annual rings.

Most of cambium cells are **fusiform initials** forming axial vessel elements, while some cambium cells are **ray initials** and they form **rays**: combinations of parenchyma cells and tracheids transporting water, minerals and sugars (because it is dark inside the stem and only respiration is possible) *horizontally*. Rays are visible best on the **tangential** section of the stem (when section plane is tangent to the stem surface); two other possible sections (**radial** and **transverse**) show axial components of the stem better. In the secondary phloem, rays are sometimes *dilated* (wedge-shaped).

The cambium usually does not work evenly all year round. In temperate climates, a *ring* forms for each growing season and makes it possible to determine the age by counting the growth rings. This is because at the end of season cambium makes much smaller ("darker") tracheary elements. Trees growing in climates without well-expressed seasons will not make annual rings. To tell the age of a tree, researchers

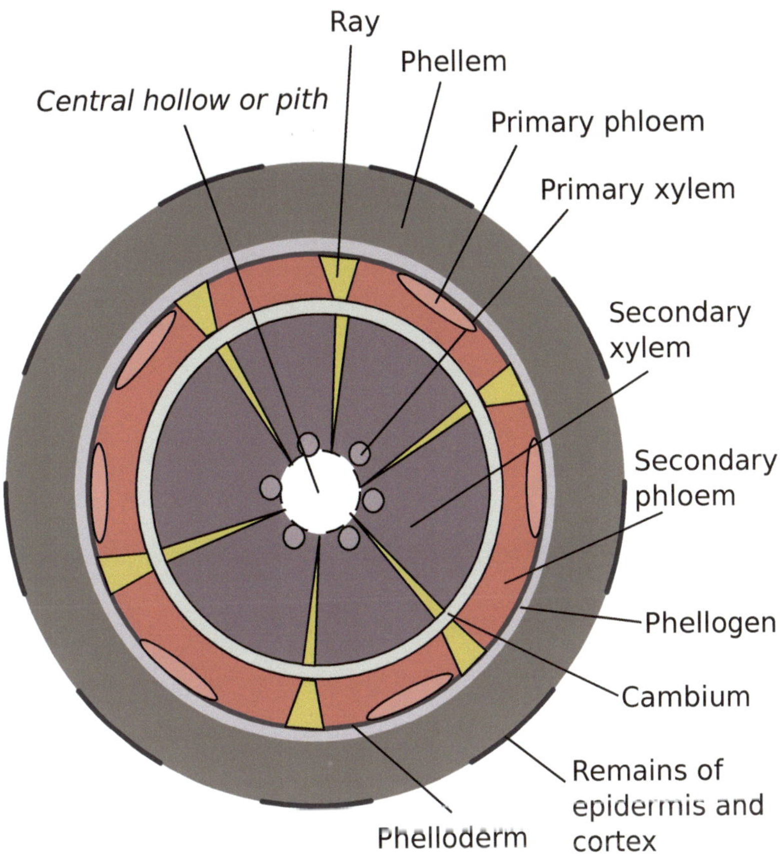

Figure 7.2. Anatomy of the secondary stem. Radial view.

observe the number and thickness of annual rings that are formed. This is called *dendrochronology*.

Some trees (like oaks, *Quercus*) have large vessel elements are found primarily in the wood formed early in the season (early wood); this pattern is known as **ring porous**. Large vessel elements of other trees (like elm, *Ulmus*) occur more evenly in both early and late wood. This pattern is known as **diffuse porous**wood: with large vessel elements in both early and late wood.

Vesselless wood of conifers is of a simpler structure with relatively few cell types. There are simple rays and frequently *resin ducts*; resin is secreted by specialized cells.

In the tree trunk, the lighter wood near the periphery is called **sapwood** and has functioning xylem where most of the water and minerals are transported. Darker wood closer to the center is called **heartwood** and is a non-functional, darkly colored

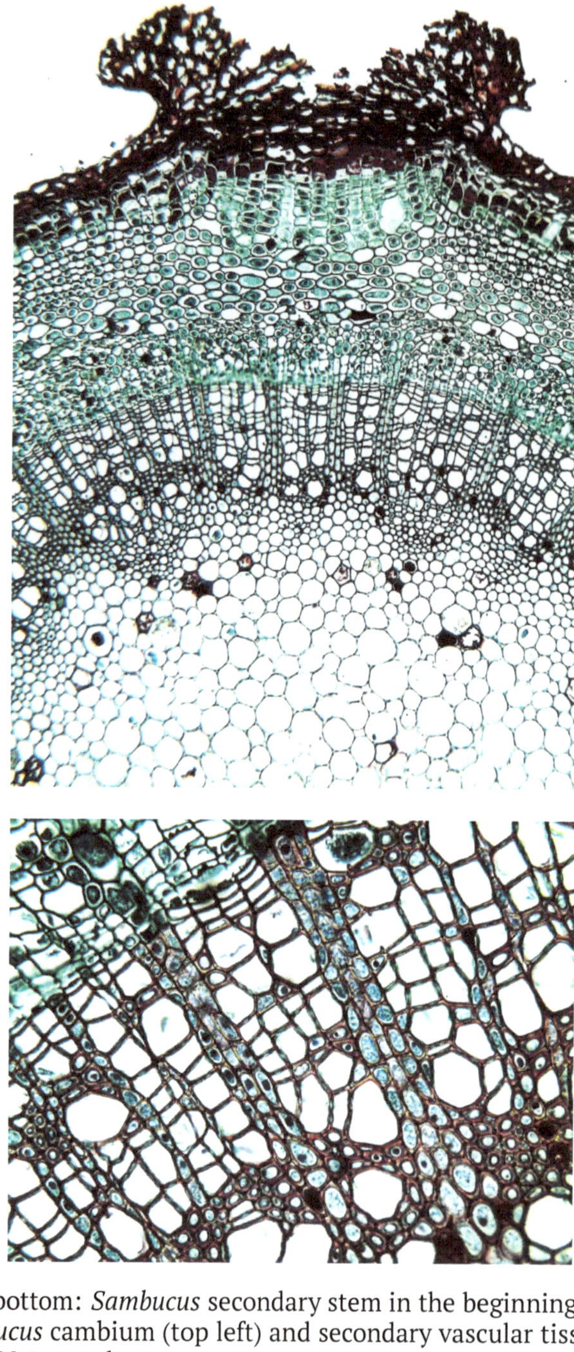

Figure 7.3. Top to bottom: *Sambucus* secondary stem in the beginning of growing, lenticel is on the top, *Sambucus* cambium (top left) and secondary vascular tissues. Magnifications $\times 100$ (first) and $\times 400$ (second).

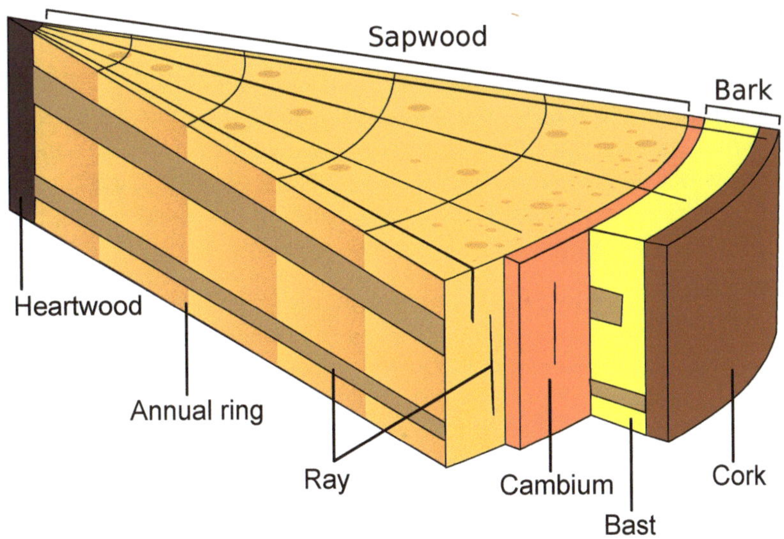

Figure 7.4. Piece of trunk. Radial and transverse views.

xylem (Fig. 7.4). Tracheary elements are dead cells and to block them, plants uses **tyloses** which also help control winter functioning of vessels. A tylose forms when a cell wall of parenchyma grows through a pit or opening into the tracheary element; they look like bubbles.

Most liliids (for example, palms) do not have lateral meristems and true wood. Some thickening does occur in a palm but this happens at the base of the tree, as a result of adventitious roots growing. Palms may also have diffuse secondary growth which is division and enlargement of some parenchyma cells. These processes do not compensate the overall growth of plant, and palms frequently are thicker on the top than on the bottom. Few other liliids (like dragon blood tree, *Dracaena*) have **anomalous secondary growth** which employs cambium but this cambium does not form the stable ring.

Constantly thickening stem requires constantly growing "new clothes", secondary dermal tissue, periderm. Periderm is a part of bark. *Bark is everything outside vascular cambium*. It is unique structure which is sometimes called "tertiary tissue" because it consists of primary and secondary tissues together:

- trunk = wood + vascular cambium ("cambium") + bark
- wood = secondary xylem + primary xylem + [pith][1]
- bark = bast (primary + secondary phloem) + periderm + [cortex] + [epidermis]

[1] "Optional" tissues are given in brackets, synonyms in parentheses.

- periderm = [phelloderm] + cork cambium (phellogen) + cork (phellem)

Each year, a new layer of phellogen (cork cambium) appears from the parenchyma cells of the secondary phloem which makes bark multi-layered and uneven. On the surface of a young stem, one may see **lenticels**, openings in phellem layer which supply the internals of the stem with oxygen; together with rays, lenticels work as ventilation shafts. To produce lenticels, some phellogen cells divide and grow much faster which will finally break the periderm open.

Apart from the lenticels, older or winter stems have **leaf scars** with **leaf traces** on their surface. The first are places where leaf petiole was attached, and the second are places where vascular bundles entered the leaf.

The secondary structure of root reminds the secondary structure of stem, and with time, these two organs become anatomically similar.

7.2 Branching Shoot

Secondary stem allows for extensive branching. In seed plants, branching is based on the axial buds. These buds are located in axils of leaves and develop into secondary shoots. There are two main types of branching: monopodial and sympodial (Fig. 7.5).

Monopodial branching is when the buds do not degrade and all the shoots continue to grow.

Sympodial branching is when the terminal buds do degrade (make FU and/or die out) and the lateral shoot closest to the terminal bud now becomes the terminal shoot and continues the vertical growth. This happens because the terminal SAM suppresses the downstream meristems by producing the auxin hormone (apical dominance). Apical dominance is a basis of multiple gardening trimming techniques.

Monopodial branching creates the conical (spruce-like) crown whereas sympodial branching will create crowns of many different shapes. Monopodial growth is considered to be more primitive. Some monopodial trees may even die if the terminal bud is damaged.

Even more ancestral mode of branching is **dichotomous**, when every branch splits into two; this is frequent in lycopods and some other Pteridophyta.

7.3 Life Forms

Thickening and branching change the appearance of plant. The most ancient classification employ both branching and thickening and divide plants into trees, shrubs

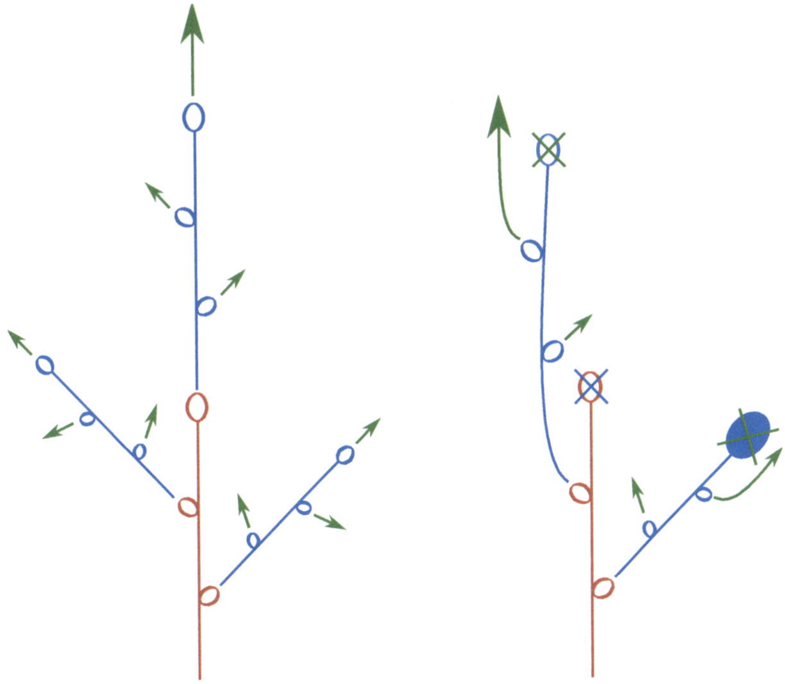

Figure 7.5. Monopodial (left) and sympodial branching. First, second and third years of growth are red, blue and green, respectively. Note that rightmost branch developed the FU (blue oval).

and herbs. This approach was the first classification of *life forms*. Life forms tell not about evolution, but about how plant lives. We still use this classification. With some modifications, it plays a significant role in gardening:

Vines Climbing woody and herbaceous plants

Trees Woody plants with one long-lived trunk

Shrubs Woody plants with multiple trunks

Herbs Herbaceous plants, with no or little secondary xylem (wood). Sometimes, divided further into annuals (live one season), biennials (two seasons) and perennials.

This classification system has many downfalls. What is, for example, the raspberry? It has woody stems but each of them lives only two years, similar to biennial herbs. Or what is duckweed? These small, water-floating plants with ovate non-differentiated bodies are hard to call "herbs". As one can see, the actual diversity of plant lifestyles is much wider than the classification above.

7.3.1 Architectural Models Approach

During the winter, it is easy to see that some tree crowns have similar principles of organization. In the winter-less climates, the diversity of these structures is even higher. On the base of branching (monopodial or sympodial), location of FU, and direction of growth (**plagiotropic**, horizontal or **orthotropic**, vertical), multiple *architectural models* were described for trees. Each model was named after a famous botanist such as Thomlinson, Corner, Attims, and others. In temperate regions, one of the most widespread models is Attims (irregular sympodial growth): birches (*Betula*) and alders (*Alnus*) grow in accordance with that model (Fig. 7.6). In tropical regions, many plants (like palms and cycads) have single thick trunks crowned with large leaves, this is Corner model (Fig. 7.7).

Figure 7.6. Attims architecture model of tree growth. (*From Halle et al. (1978)*).

7.4 Modified Shoot

Like leaves and roots, shoots and stems also have modifications. Some examples are rhizomes, stolons, tubers, bulbs, corms, thorns, spines, cladophylls, and stem traps. **Rhizomes** (example: ginger, *Zingiber*) are underground stems that burrow into the ground just below the soil surface, and usually tend to have small, scale-like leaves that are not photosynthetic. Buds from the axils of the leaves make new branches that will grow to become aboveground shoots. **Stolons** (runners) are aboveground horizontal shoots, which sprout and produce a new plants (example: strawberry, *Fragaria*). **Tubers** (example: potatoes, *Solanum*) are enlarged portions of rhizomes. The "eyes" of potato are actually lateral buds and the tuber body is comprised of many parenchyma cells that contain amyloplasts with starch. Corms and bulbs are shoot

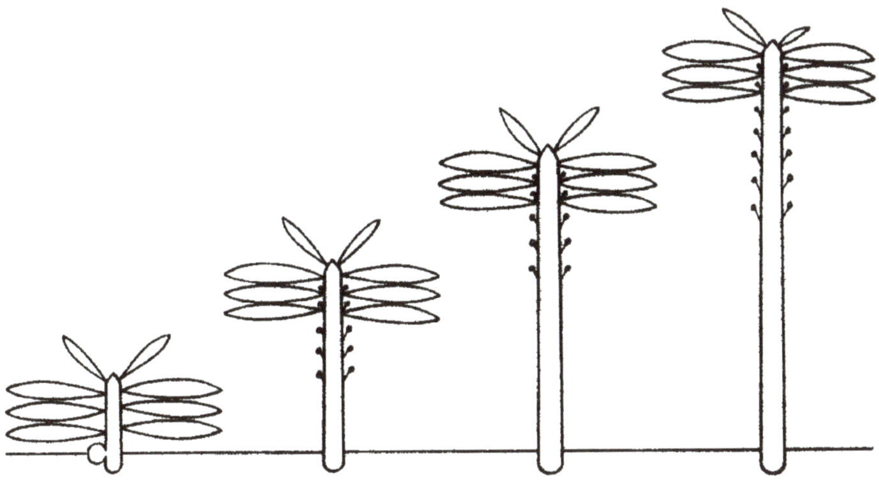

Figure 7.7. Corner architecture model of tree growth. (*From Halle et al. (1978)*).

structures that are used for storage. A **corm** (example: crocus, *Crocus*) is a short, thick underground storage stem with thin scaly leaves. A **bulb** (example: onion, *Allium*) differs from a corm in the fact that it stores its nutrients in its fleshy leaves (Fig. 7.8).

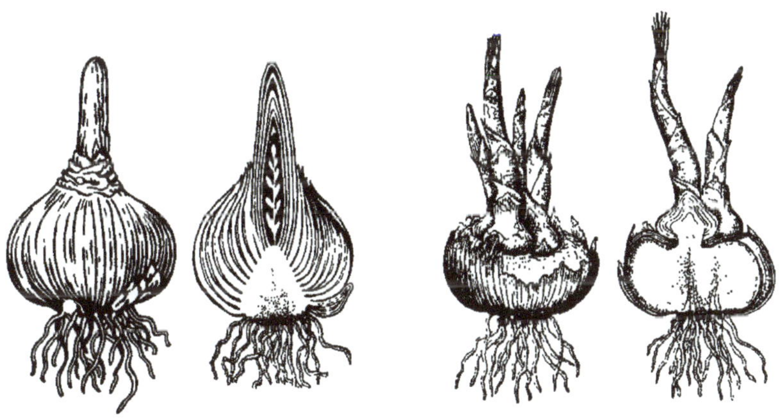

Figure 7.8. Bulbs (left) and corms. (*Modified from various sources*).

Thorns (example: hawthorn, *Crataegus*) are defensive shoots that help to protect the plant from predators. **Spines** are not modified stems, but rather modified, reduced leaves or stipules, or bud scales (example: almost all cacti, Cactaceae family). **Prickles** (example: rose, *Rosa*) are modified surface tissues of stem.

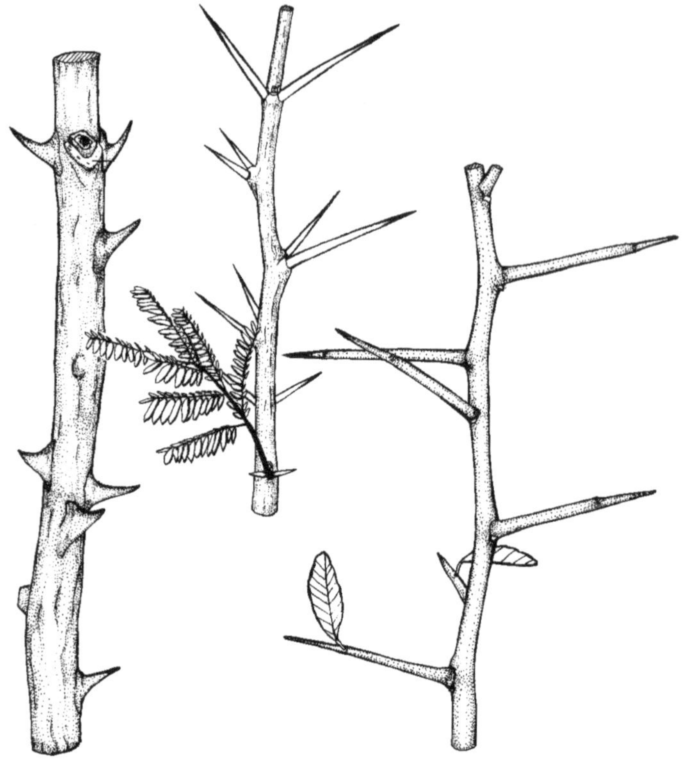

Figure 7.9. Prickles, spines (from stipules) and thorns (from Charles-Dominique et al., 2016).

Cladophylls (examples: Christmas cactus, *Schlumbergera*; ribbon plant, *Homalocladium*) are leaf-like, flattened shoots. **Phyllodes** are actually leaf modifications (example: Australian acacias, *Acacia*) they visually similar to cladophylls but originated from flattened leaf petioles. Shoot **insect traps** are used by some carnivorous plants, such as bladderwort (*Utricularia*). The following table emphasizes the diversity of organ modifications:

Function \ Organ	Leaf	Stem/shoot	Root
Absorption	Absorption leaves (bromeliads)	Rhizoids	*Default*
Defense	Spines, scales	Thorns, prickles	Spines
Expansion	Plantlets	Rhizomes, stolons, runners	Adventive buds
Interactions	Traps, sticky epidermis, urns, colored leaves	Traps, insect nests	Haustoria, mycorrhizae, root nodules, nematode traps, insect nests
Photosynthesis	*Default*, phyllodes	Cladophylls	Green roots (orchids)
Storage	Succulent leaves, pitchers	Bulbs, corms, tubers	Storage roots
Support	Tendrils, false stems, floats, suckers	*Default*, tendrils	Buttress, aerial and contractile roots, suckers

Please note that superficially similar structures (e.g., shoot and leaf tendrils) might have different origin.

7.4.1 Raunkiaer's Approach

Christen Raunkiaer used a different approach to classify life forms which is useful to characterize the whole *floras* (all plant species growing on some territory), especially temperate floras. He broke plants down into six categories: epiphytes, phanerophytes, chamaephytes, hemicryptophytes, cryptophytes and therophytes.

Epiphytes do not touch soil (they are aerial plants), *phanerophytes* have their winter buds exposed, *chamaephytes* "put" their winter buds under the snow, winter buds of *hemicryptophytes* on the soil surface, *cryptophytes* in the soil and/or under water, and *therophytes* do not have winter buds, they go through winter as seeds or vegetative fragments (Fig. 7.10)[2]. Typically, northern floras have more plants of last categories

[2]They also distinguish aerophytes, "aerial plants".

whereas first categories will dominate southern floras. Note that Raunkiaer "bud exposure" is not far from the hardiness in the dynamic approach explained below.

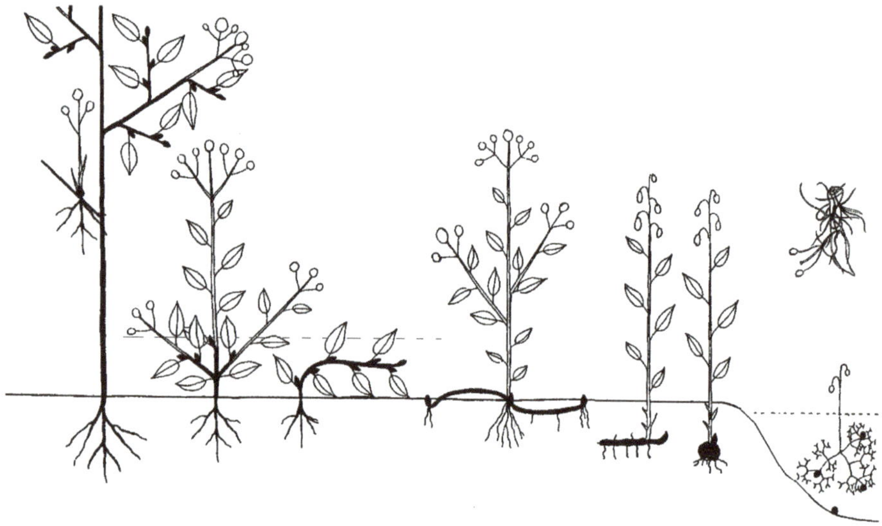

Figure 7.10. Raunkiaer life forms. From left to right: epiphyte (on branch), phanerophyte, chamaephyte, hemicryptophyte, two cryptophytes (with rhizome and with bulb), therophyte (in water) and aerophyte (in air). Dashed line on the left is the projected snow level. *(From Raunkiaer (1907), extended)*.

7.4.2 Dynamic Approach

There are many life forms classifications. This is because life forms represent numerous *secondary patterns* in plan diversity, along with the main pattern which taxonomy wants to describe.

Dynamic life forms classification uses the fact that in nature, there are no strict borders between different life forms. If we supply the pole to some shrubs, they may start to climb and therefore become vines. In colder regions, trees frequently lose their trunks due to low temperatures and form multiple short-living trunks: they become shrubs. Conversely, in tropics, many plants which are herbs in temperate regions, will have time to develop secondary tissues and may even become tree-like.

Dynamic approach uses three categories: hardiness, woodiness, and slenderness (Fig. 7.11). *Hardiness* is a sensitivity of their exposed parts to all negative influences (cold, heat, pests etc.) This is reflected in the level of plant exposure, plants which are hardy will expose themselves much better. *Woodiness* is the ability to make dead tissues, both primary and secondary (reflected in the percentage of cells with secondary walls). High woodiness means that plants will be able to support themselves without problems. *Slenderness* is an ability to grow in length (reflected in the proportion of linear, longer than wide, stems). Low slenderness results in rosette-like plants. Combining these three categories in different proportions, one may receive all possible life forms of plants.

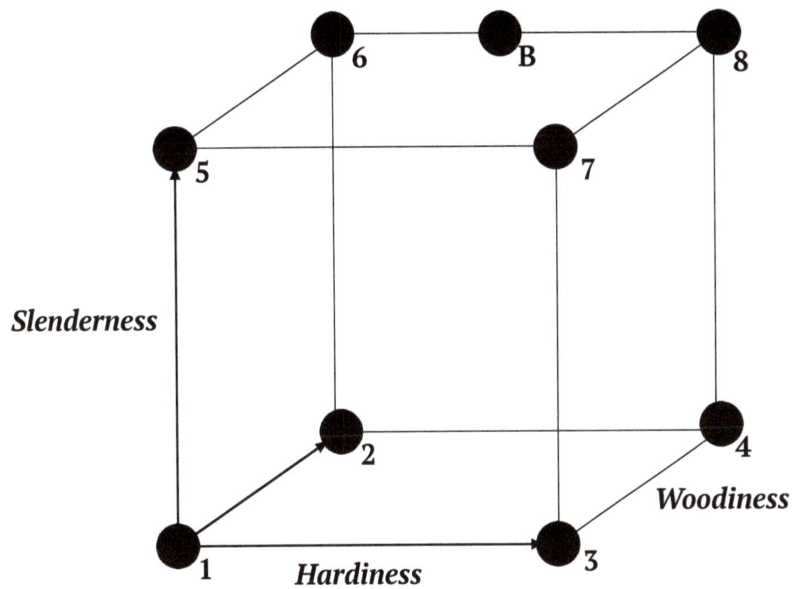

Figure 7.11. Dynamic life forms: 3D morphospace.

These three categories could be used as variables of the 3D morphospace. Every numbered corner in the morphospace diagram (Fig. 7.11) represents one extreme life form:

1. Reduced floating annuals like duckweed (*Lemna*). Please note that zero hardiness is impossible; duckweed hardiness is just low.
2. Short annual herbs like marigold (*Tagetes*); they accumulate wood if warm season is long enough.
3. Bulb perennials like autumn crocus (*Colchicum*).
4. Australian "grass trees" (*Xanthorrhea*) with almost no stem but long life.
5. Herbaceous vines like hops (*Humulus*).
6. Monocarpic tree-like plants like mezcal agave (*Agave*).
7. Perennial ground-cover herbaceous plants like wild ginger (*Asarum*).
8. Trees like redwood (*Sequoia*).

What is even more important, all possible positions on the "surface" and inside this cube also represent life forms. For example, the dot marked with "B" are slender, woody but only partly hardy plants. The partial hardiness means that vertical axes will frequently die, and then new slender woody axes develop from scratch. Woody wines and creeping bushes will correspond well with this description. As you see, this morphospace not only classifies existing plants but also could predict possible life forms.

7.5 Origin of the Seed

When plants developed the secondary growth, the almost unlimited perspectives opened for enlarging their body. However, these giants faced a new problem.

Big animals like elephants, lions, and whales tend to produce minimal number of offspring but increase the child care to ensure survival. This is called ***K*-strategy**, this is opposite to ***r*-strategy** of usually smaller creatures which employs big numbers of offspring, and most of them will not survive (Fig. 7.12).

Figure 7.12. K-strategic elephant and r-strategic (like codfish) elephant. The second does not exist. Why?

Analogously, bigger plants would need to do the same as K-strategic animals: make few daughter plants but defend and supply them with all needs until they mature. However, big secondary thickening spore plants were not capable of family planning; they still made billions of spores and then left them to fend for themselves. Naturally, only few from these billions would survive to become fertilized.

Spore reproduction is cheap and efficient but as birth control is not available, results are unpredictable. Even worse, these spore tree forests were not at all stable: in accidentally good conditions, many spores would survive and make sporophytes which start to grow simultaneously and then suppress each other and even die from over-population. But if the environmental conditions are bad, then none of the gametophytes will survive so there would be no new saplings to replace the old trees.

It is similar to the so-called "dinosaur problem". This situation arose when giant Mesozoic reptiles also lost the control for their offspring: their egg size was limited due to physical restrictions, therefore, young dinosaurs were so much smaller than adults; then the only possible strategy was to leave them alone (Fig. 7.13). As a result, at the end of Mesozoic dinosaurs either decreased in size (became birds), or went extinct.

Plants, however, kept their size and survived. This is because they developed the seed (Fig. 7.14).

A *seed is the result of enforced control of the sporophyte over the gametophyte*. The idea of a seed is to hide most of the heterosporous life cycle *inside mother plant* (Fig. 7.15). In seed plants, everything happens directly on the mother sporophyte: growing of gametophytes, syngamy, and growing of daughter sporophyte. Consequently, the female spore (megaspore) never leaves the sporangium. It germinates inside, waits for fertilization and then the zygote grows into and embryo, still inside the same sporangium.

What will finally leave the mother plant is the *whole female sporangium with gametophyte and embryo on it*. This is the **seed**. It can be defined as *chimeric structure with three genotypes*: seed coat (mother plant megasporangium, $2n$), endosperm (female gametophyte, n), and daughter sporophyte (embryo, $2n$).

It should be noted here that flowering plants have endosperm of different origin; it is called *endosperm*$_2$ and usually is triploid ($3n$) whereas female gametophyte endosperm is haploid (n) **endosperm**$_1$. The other note is that apart from seed coat (which originates from **integument**(s), megasporangium extra cover(s)), mother sporophyte also gives **nucellus** (wall of megasporangium) which sometimes is used as a feeding tissue for the embryo. This last tissue is called **perisperm**.

Figure 7.13. How dinosaurs and ferns solved the problem of conflicting strategies.

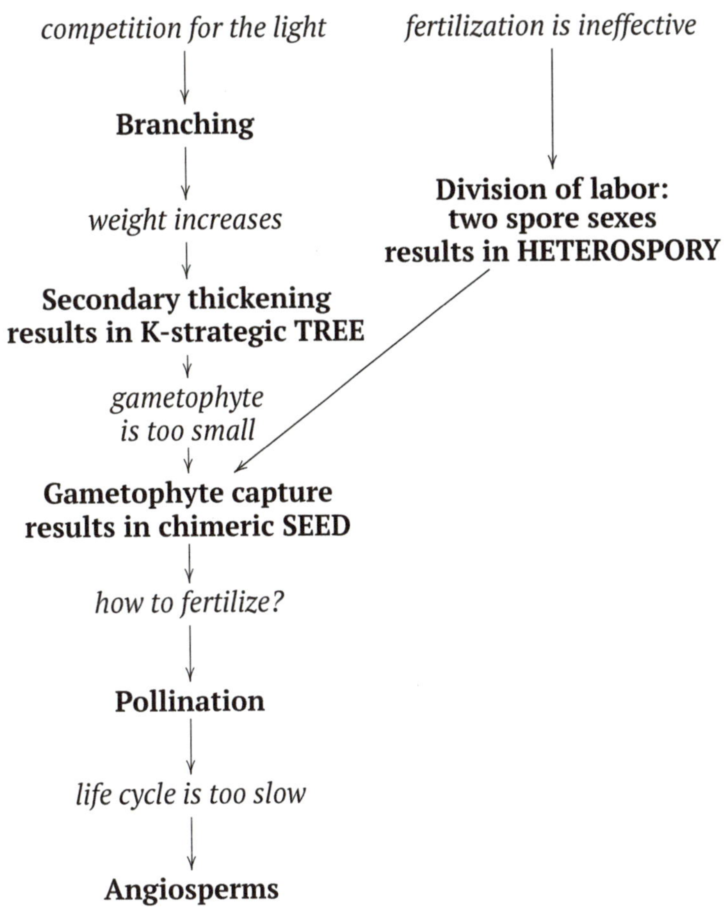

Figure 7.14. How secondary thickening, heterospory and seed are connected: *challenges* to land plants and their **responses**. This is part 2, part 1 is on Fig. 5.3.

One problem is still left. How will sperms reach female gametophyte and egg cell? The target is now high above the ground, on a branch of the giant tree. The only possible solution is pollination.

Pollination is the distribution of the *whole male gametophytes* which are called **pollen grains**. Plants have no legs so they always need a third party in their sex, this is mostly wind or insects.

A pollen grain is *not a spore*, mother sporophyte cares about male lineage too, and male spore grows into very small male gametophyte. It contains multiple haploid cells; some of which are sperms.

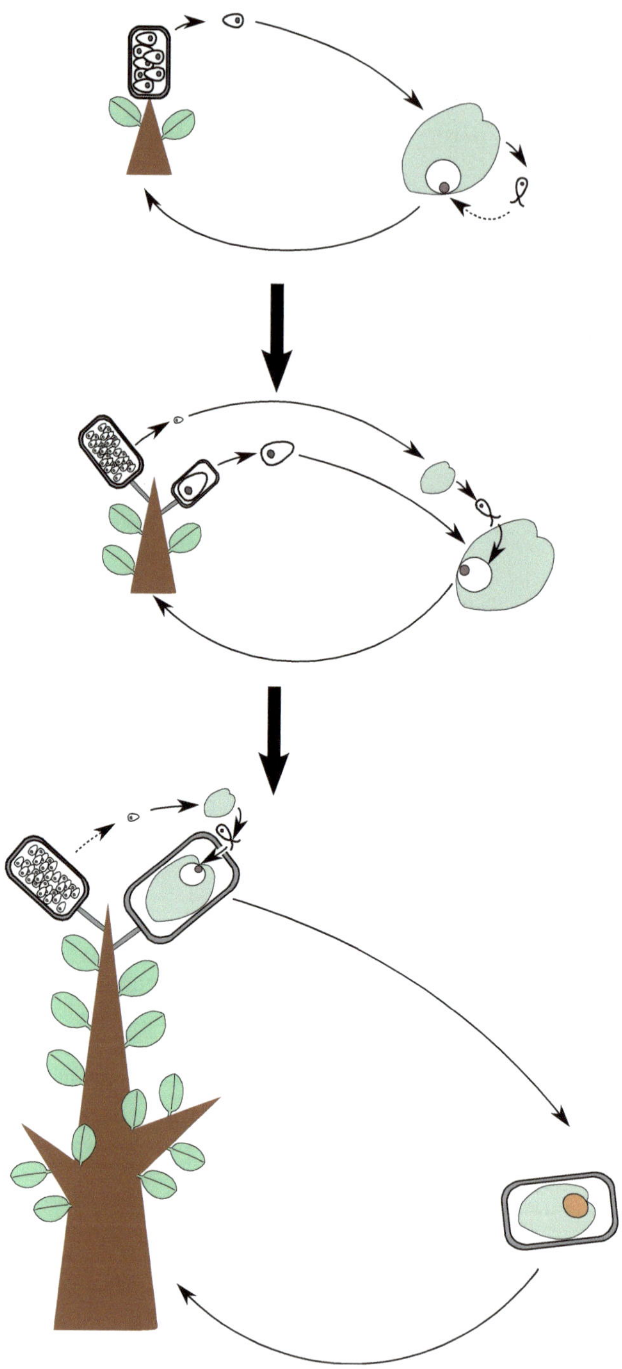

Figure 7.15. The origin of seed (see explanation in the text). Sporophyte is on the left, gametophytes and seed—on the right or/and on the top. First stage—isosporic life cycle, second stage—heterosporic life cycle, third stage—seed plant life cycle.

The lesser problem is: How would these sperms will swim to the egg cell? Some seed plants will excrete the drop of liquid from the top of the **ovule** (integument(s) + megasporangium), whereas the other, more advanced way is to grow a sperm delivery tool, the **pollen tube** (Fig. 7.16) made from one of the pollen grain cells. Fertilization with pollen tube is often called **siphonogamy**.

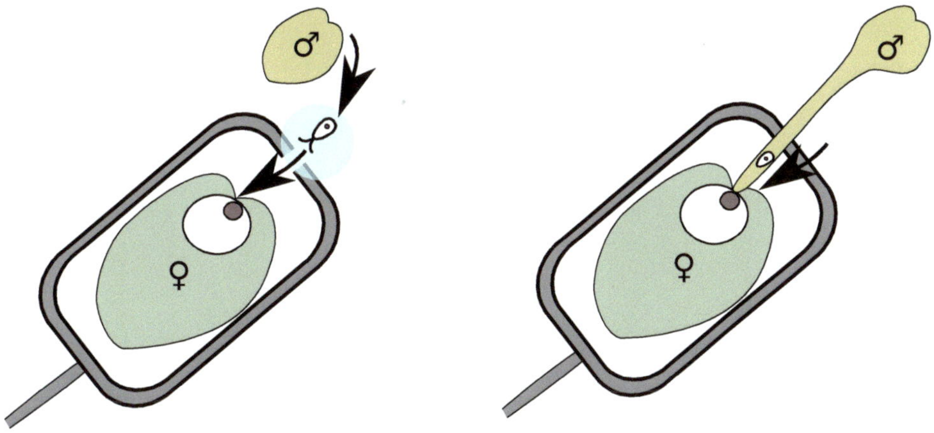

Figure 7.16. Open fertilization in seed plants (left) with the help of liquid extracted by mother plant; and fertilization with pollen tube (right), or *siphonogamy*, with the help of pollen tube growing from male gametophyte. Among extant seed plants, only cycads and ginkgo have the open fertilization.

Consequently, seed plants with the pollen tube do not have flagella even on male gametes; these cells are **spermatia**: aflagellate, non-motile male gametes. (Below, we will continue to call all male gametes "sperms"). Pollen tube also allows only two male gametes per gametophyte: in living world, male gametes are usually competing for fertilization—this selects the best genotypes; whereas in higher seed plants, competition is between pollen tubes. Haploid pollen tube grows inside alien tissue of diploid sporophyte, so this growth is extremely slow in many seed plants. However, angiosperms made their pollen tubes grow fast.

With all these revolutionary adaptations, seed plants were first to colonize really dry places, and, in turn, allowed all other life to survive in arid climates.

The cycle of a seed plant (Fig. 7.17) begins with a sporophyte ($2n$) and has both the female and male organs where some cells undergo meiosis. Inside the ovule (which is the megasporangium with extra covers), female gametophyte (n, future endosperm$_1$) produces the egg cells. Male gametophytes (pollen grains) ripen in the **pollen sac** which is the microsporangium. The pollen sac sends out the pollen grains which meet up with the ovule. The pollen grain then releases the sperms which fertilize the egg cell, and a zygote is formed. The zygote grows into embryo (which uses endosperm as a feeding tissue) and then into the sporophyte.

Several plant lineages met this "seed challenge", there were seed lycophytes and also seed "horsetails". However, seed ferns made it first and became ancestors of *seed plants*.

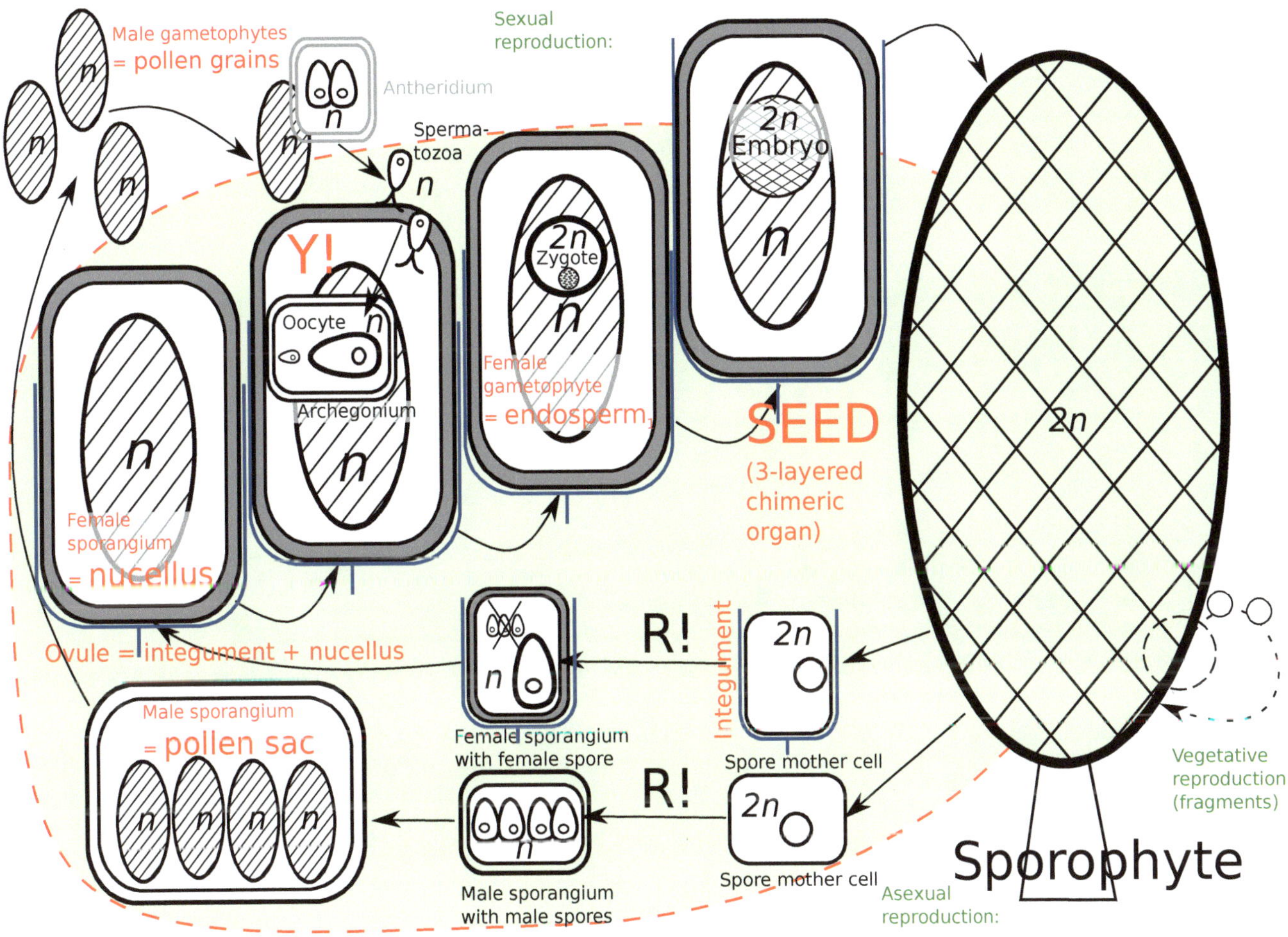

Figure 7.17. Life cycle of seed plants. Innovations (comparing with the heterosporic life cycle) are in red.

7.5.1 Seed Structure and Germination

Seeds are diverse. For example, in an onion (*Allium*), a seed (Fig. 7.18) has endosperm, one **cotyledon** (embryonic leaf), *radicle* (embryonic root), and the lateral embryonic bud (*plumula*).

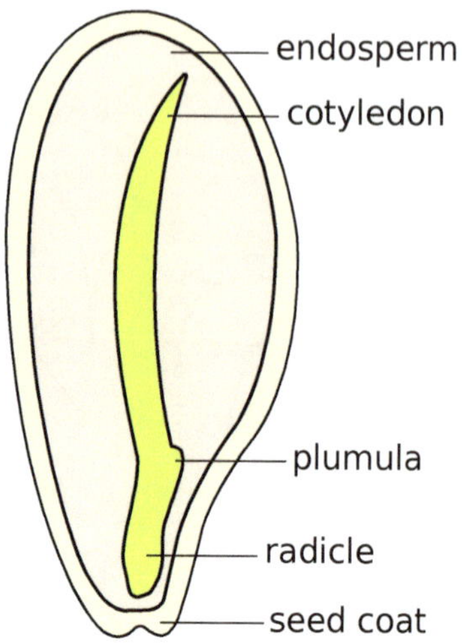

Figure 7.18. Monocot seed.

Beans (*Phaseolus*) and other Leguminosae are examples of seeds without endosperm—actually, it was there, but growing embryo usually eats it out completely. These seeds have two large cotyledons. Grass (Gramineae) seeds contain several specific organs, namely coleoptile, coleorhiza, and scutellum. The *scutellum* is an enlarged cotyledon, *coleoptile* is the bud cover, and *coleorhiza* covers the embryonic root, radicle (Fig. 7.19). Onion and grasses are monocots with lateral embryonic bud. Other seed plants have a terminal embryonic bud and two or multiple cotyledons. Pine (*Pinus*) is an example of a plant that has multiple (five or more) cotyledons. Some plants like orchids (Orchidaceae) do not have developed embryo and even endosperm in seeds, their germination depends on a presence of symbiotic (mycorrhizal) fungus.

The first step in germination and starts with the uptake of water, also known as *imbition*. After imbition, enzymes are activated that start to break down starch into sugars consumed by embryo. The first indication that germination has begun is a swelling in the radicle. In onion and pea (*Pisum*), a structure that looks like a hook goes up through the soil and expose cotyledons and both hypocotyl and epicotyl (first internode). In beans, grasses, and palms, only epicotyl is exposed aboveground whereas cotyledons and hypocotyl remain underground.

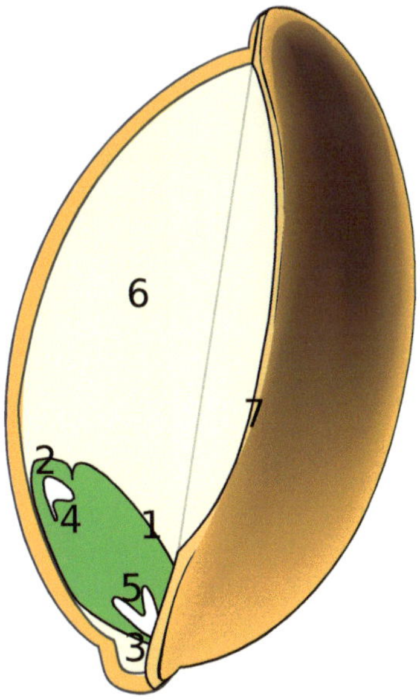

Figure 7.19. Grass seed. 1 scutellum (= cotyledon), 2 coleoptile (bud cover), 3 coleorhiza (radicle cover), 4 embryo bud (= plumula), 5 radicle (= embryo root), 6 endosperm, 7 seed coat.

7.6 Spermatophyta: seed plants

Seed plants consist of approximately 1,000 species of non-angiosperms (gymnosperms) and about 250,000 species of angiosperms. They have a sporic life-cycle with sporophyte predominance, and seeds. The gametophyte is reduced to cells inside the ovule or pollen grain. Males have a minimum number of cells being three and females being four. The antheridia are absent and in flowering plants (Angiospermae) and Gnetopsida the archegonia are also reduced. The sporophyte will always start as an embryo located inside the nutrition tissue, $endosperm_1$ which is the female gametophyte or in $endosperm_2$ (see the next chapter). Spermatophyta have axillary buds (buds in leaf axils). Like ferns, they are megaphyllous and homoiohydric, and have a secondary thickening. Higher groups of seed plants lost flagellate spermatozoa and developed pollen tubes. The classes of Spermatophyta are Ginkgoopsida, Cycadopsida, Pinopsida, Gnetopsida, and Angiospermae.

Ginkgoopsida is just one species; ginkgo or maidenhair tree (*Ginkgo biloba*). This plant is long extinct in the wild but is grown on Chinese temple grounds as a decorative tree. Ginkgo is a large tree bearing distinctive triangle-shaped leaf with dichotomous venation. This plant is also dioecious (as an exception among plants, *Ginkgo* has sexual chromosomes like birds and mammals) and the pollen is transported by wind to female (ovulate) trees. The pollen grains of the ginkgo plants produce two multi-flagellate spermatozoa; the edible seed is fruit-like and becomes ripe after lying on the ground for a long time. Maidenhair tree has symbiotic cyanobacteria in cells. As ginkgo probably went through the population bottleneck, there are very few, almost no, phytophagous insects that can damage ginkgo leaves. The only fungus which is capable to eat them, *Bartheletia*, is also a living fossil.

Cycadopsida—cycads is a class with few genera and about 300 species that grow mostly in tropics. Only one species grows naturally in the United States, *Zamia pumil*a, and can be found in Florida and Georgia. Cycads are palm-like plants with large, pinnate leaves. Their wood is rich of parenchyma since stem has anomalous secondary thickening. They are all dioecious and its cone is large and protected by prickles and woody plates. The ovules of these plants are attached to modified leaves (**megasporophylls**) that are gathered in upright cones. Like ginkgo, they have multi-flagellate spermatozoa, archegonia and large oocyte. Cycad seeds are distributed by animals. Life cycle is extremely slow.

Pinopsida—conifers are the most widely known and economically important among gymnosperms. Conifers consist of approximately 630 species. Most of them are temperate evergreen trees, but some are deciduous, such as larch (*Larix*). The stem has a large amount of xylem, a small cork, and minute pith. The ovules are attached to specialized leaves, **seed scales**, and are compacted in cones with **bract scales** (Fig. 7.20). Some conifers, like junipers (*Juniperus*) and yews (*Taxus*), lack woody cones; these plants have fleshy scales. Seeds are distributed by wind and animals.

In all, conifer life cycle takes up to two years. Conifers do not have flagellate spermatozoa; their non-motile male gametes (spermatia) move inside long, fast-growing pollen tube. Among families of conifers, Pinaceae (pine family) have resin and needle-like leaves; *Pinus* have them in shortened shoots, **brachyblasts**, and their large cones have woody scales. Cupressaceae (cypress family) do not have resin, produce small cones that have a fused bract and seed scales, have dimorphic leaves, and some of their genera (like "living fossil" *Metasequoia* from China) are deciduous in an unusual way: they drop whole branches, not individual leaves.

Gnetopsida—gnetophytes are sometimes called chlamydosperms. They are a small class with only three genera that are not at all similar: *Ephedra*, *Welwitschia*, and *Gnetum*. While these plants morphologically remind of angiosperms, they are molecularly related more to other gymnosperms. *Ephedra* are horsetail-like desert leafless shrubs, *Gnetum* are tropical trees, and *Welwitschia* are plants which have a life form that is really hard to tell (Fig. 7.21).

Ephedra has archegonia, but in *Gnetum* and *Welwitschia* they are reduced. On the other hand, *Ephedra* and *Gnetum* have **double fertilization**: both male nuclei fuse with cells of the one female gametophyte (endosperm$_1$): with egg cell and another

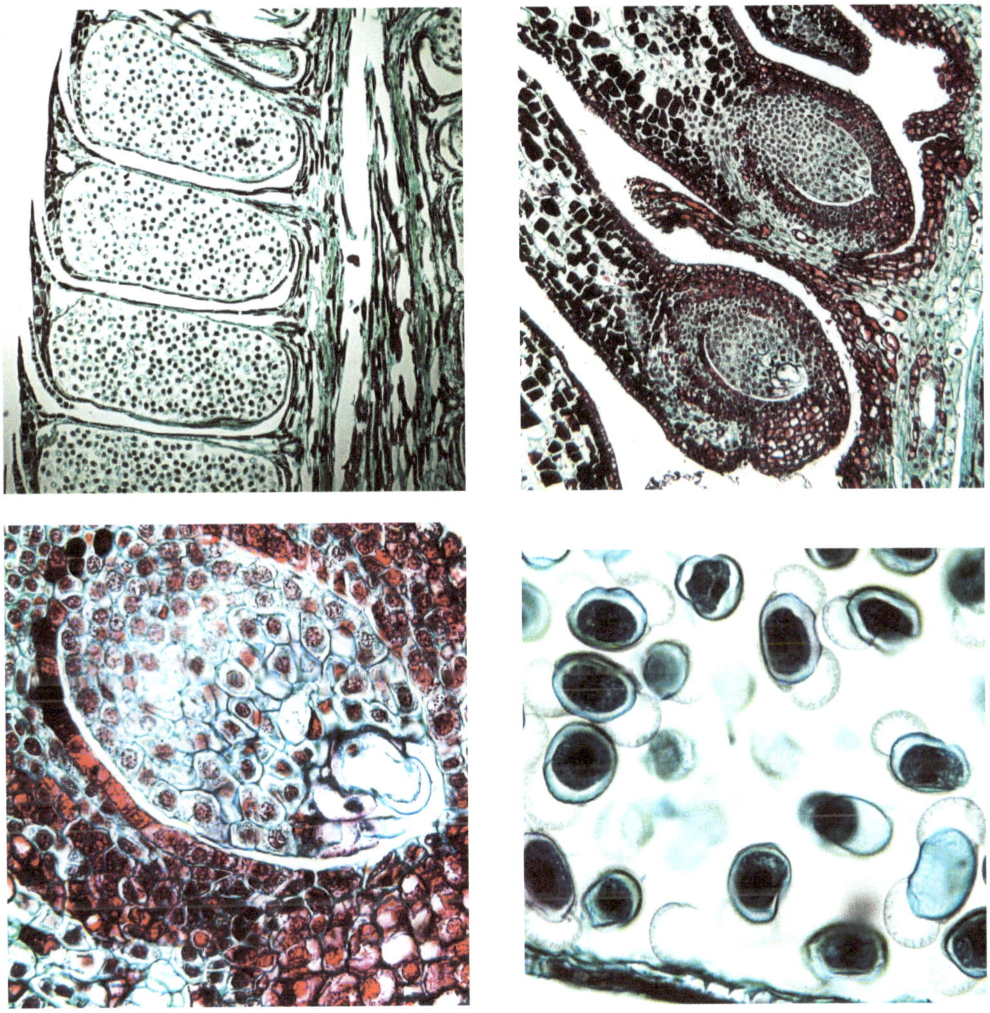

Figure 7.20. Left to right, top to bottom: *Pinus* (Pinopsida) microsporangia (pollen sacs), ovules sitting on seed scales, female gametophyte (endosperm$_1$) and multicellular male gametophyte (pollen). Magnifications $\times 100$ (first and second) and $\times 400$ (third and fourth).

haploid cell, sister to the egg. Double fertilization in gnetophytes results in two competing embryos, and only one of them will survive in future seed.

Both *Gnetum* and *Welwitschia* have vessels (like angiosperms). *Gnetum* also has angiosperm-like opposite leaves with pterodromous venation, like, for example, coffee tree (however, this probably is a result of modification of dichotomous venation). Ovules of chlamydosperms are solitary and covered with an additional outer integument; the male gametes are spermatia moving inside pollen tube.

Figure 7.21. Gnetopsida "even man out" game: *Gnetum*, *Welwitschia*, *Ephedra*, and ... *Coffee*. Which is where?

Welwitschia is probably most outstanding among gnetophytes. There is only one species that occurs in the Namibian desert. The best way to describe this plant is an "overgrown seedling." It has a small trunk with two wide leaves that have parallelodromous venation. The secondary thickening is anomalous, wood has vessels. Plant is insect-pollinated, and its winged seeds are dispersed by the wind. Fertilization is not double, but, along with pollen tubes, involves the most crazy structures: *prothallial tubes* which grow from female gametophyte and meet with pollen tubes to make zygote.

* * *

Life cycles determine the basic diversity of plants, they designate plant phyla. Let us compare three types of life cycles again (Fig. 7.22) and again (Fig. 7.23). What is visible on all these schemes, as well as on all similar schemes from above, is growing complexity of cycle, growing reduction of haploid stage, and growing self-similarity within the cycle.

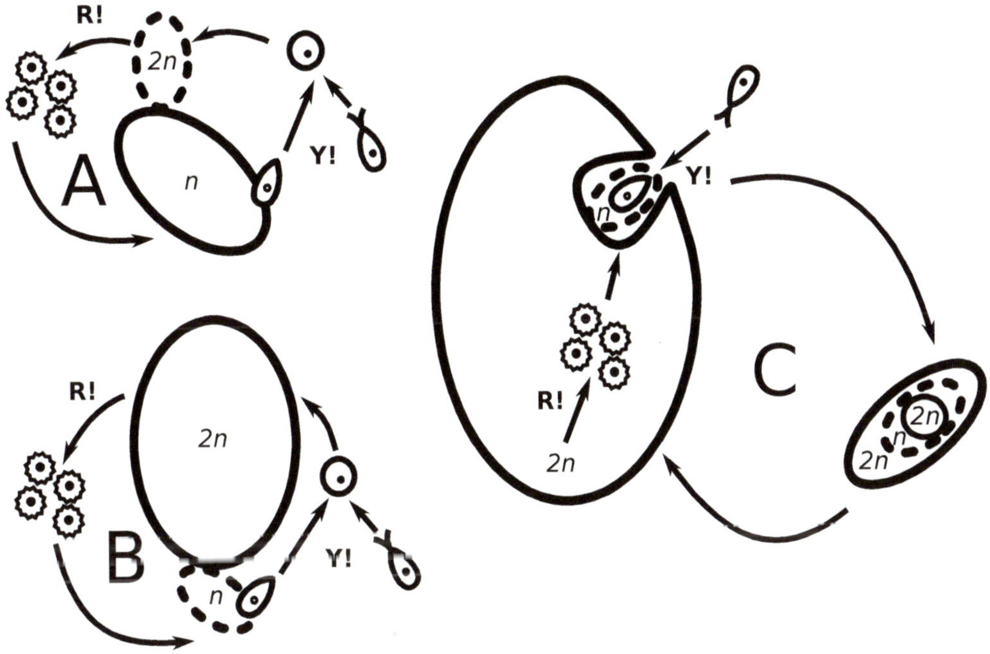

Figure 7.22. Life cycles of mosses (A), ferns (B) and seed plants (C): black and white scheme.

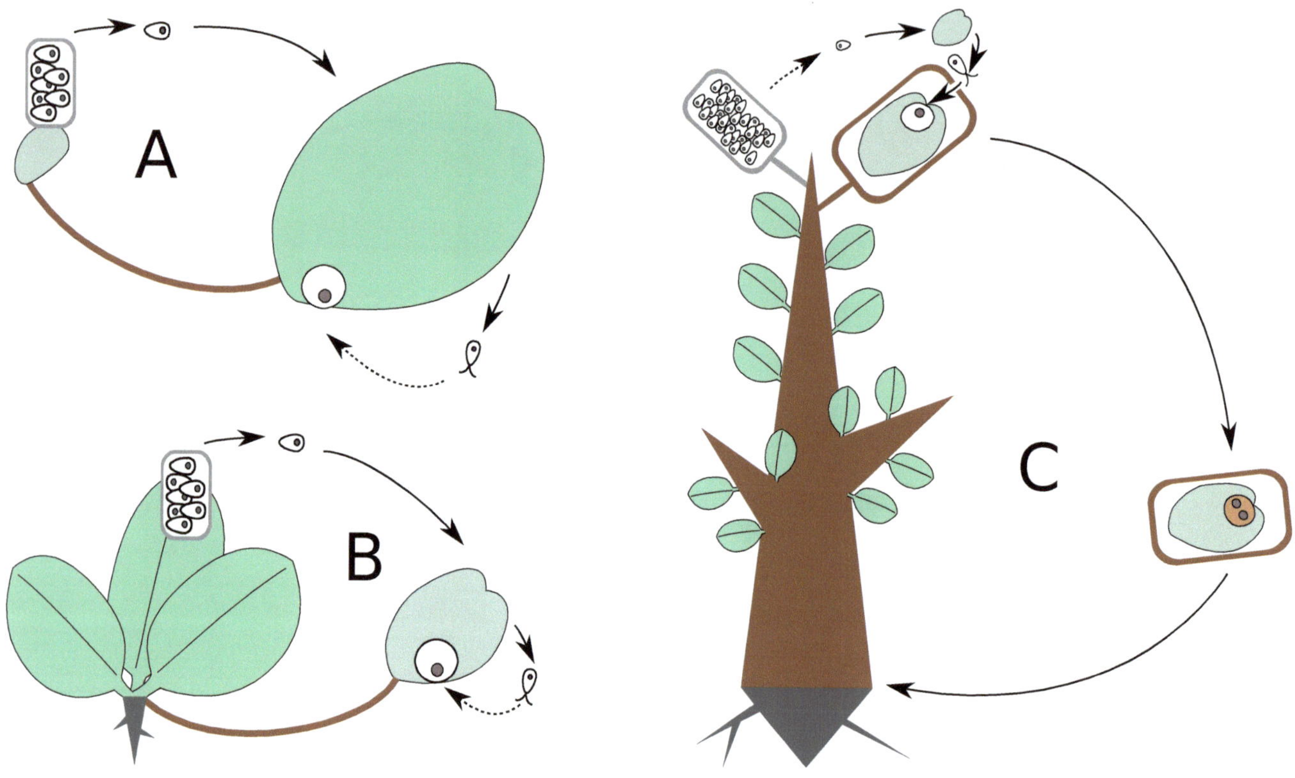

Figure 7.23. Life cycles of mosses (A), ferns (B) and seed plants (C): color scheme.

Chapter 8

The Origin of Flowering

8.1 Spermatophyta 2.0

Flowering plants (angiosperms, Angiospermae) are sometimes referred to as "Spermatophyta 2.0.", or "upgraded gymnosperms". In fact, there is no single character which unequivocally differs flowering plants from other seed plants. Only several characteristics combined together will distinguish angiosperms. Flowering plants have their ovules inside an additional cover: **pistil** which corresponds with megasporophyll (sporangium-bearing leaf); later, the pistil develops into the *fruit*. These plants have an almost complete reduction of gametophytes: three or even two cell of the pollen (male gametophyte) and seven (sometimes even four) cells in **embryo sac** (female gametophyte), there are no archegonia or antheridia. Like gnetophytes, they have double fertilization. The sperms (spermatia) come through the pollen tube (like in conifers and gnetophytes). One sperm fertilizes the egg cell, and the other sperm fertilizes the biggest cell of embryo sac (Fig. 8.1).

While the first fertilization results in a "normal" diploid zygote which grows into embryo, the second fertilization ignites the process of feeding tissue development. This feeding tissue is **endosperm$_2$**, frequently triploid ($3n$) since it *originates from the sperm and cell with two nuclei* and sperm, or diploid ($2n$), if the biggest cell of embryo sac (**central cell**) had one nucleus only.

Double fertilization may be explained in several ways:

1. the second fertilization results in second, "altruistic" embryo which sacrifices itself to feed the sibling;
2. second fertilization is only a signal which initiates the development of endosperm and it does not really matter which genotype it has;

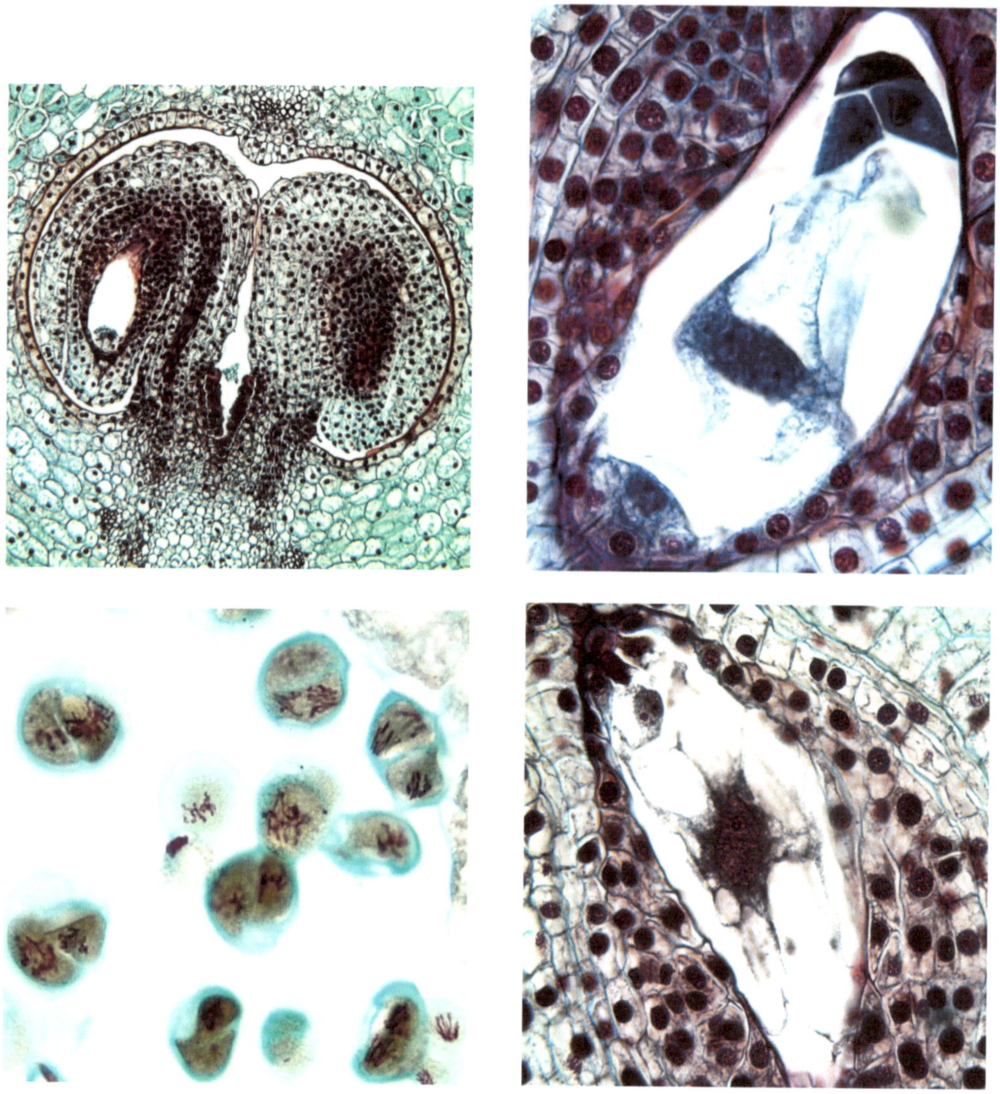

Figure 8.1. Left to right, top to bottom: *Lilium* (Liliidae) ovules, female gametophyte (embryo sac), meiosis II in pollen sacs and double fertilization (egg cell on top is fusing with first sperm, second sperm nucleus in the center is fusing with the nucleus of the central cell). Magnifications $\times 100$ (first) and $\times 400$ (others).

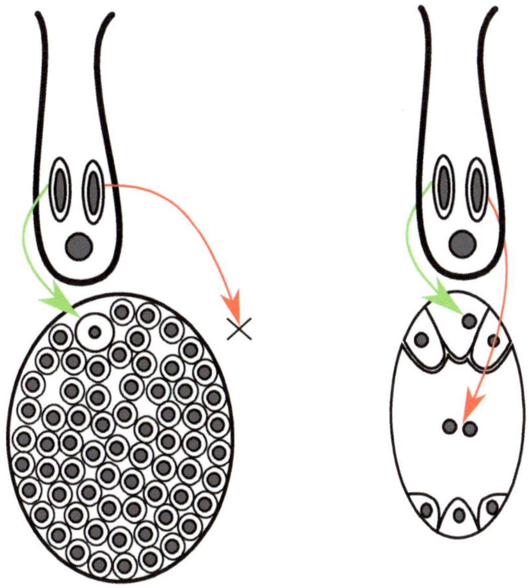

Figure 8.2. Single fertilization (left) and double fertilization (right). Egg fertilization is in green color. Note that in double fertilization second sperm fertilizes (red arrow) cell *sister to egg* (central cell in angiosperms), originated from the *same megaspore*.

3. to make a functional nutrition tissue, angiosperms need a polyploid genome whereas its origin is not so important.

Second hypothesis explains well how angiosperms saved time and resources. Third hypothesis is indirectly supported by the fact that in animals, namely two families of scale insects, there is a similar process (zygote descendant joins sister cell of the egg) which resulted in special polyploid bacteriome, tissue rich of symbiotic bacteria.

One way or another, flowering plants abandoned pre-fertilization development of the nutrition tissue, and changed endosperm$_1$ to endosperm$_2$ (Fig. 8.2).

* * *

In the Mesozoic era, gymnosperms were the dominating plants of the tree story. However, in the uderstorey, herbaceous spore plants did not surrender to seed plants and were still dominating. Amazingly, there were almost no herbaceous gymnosperms! The explanation is that gymnosperms, being quite advanced in general, had a slow and ineffective life cycle.

While ferns and mosses have one "gunshot" in their life cycles (this is fertilization, because dissemination of spores is mainly random), seed plants have *two*: first, they

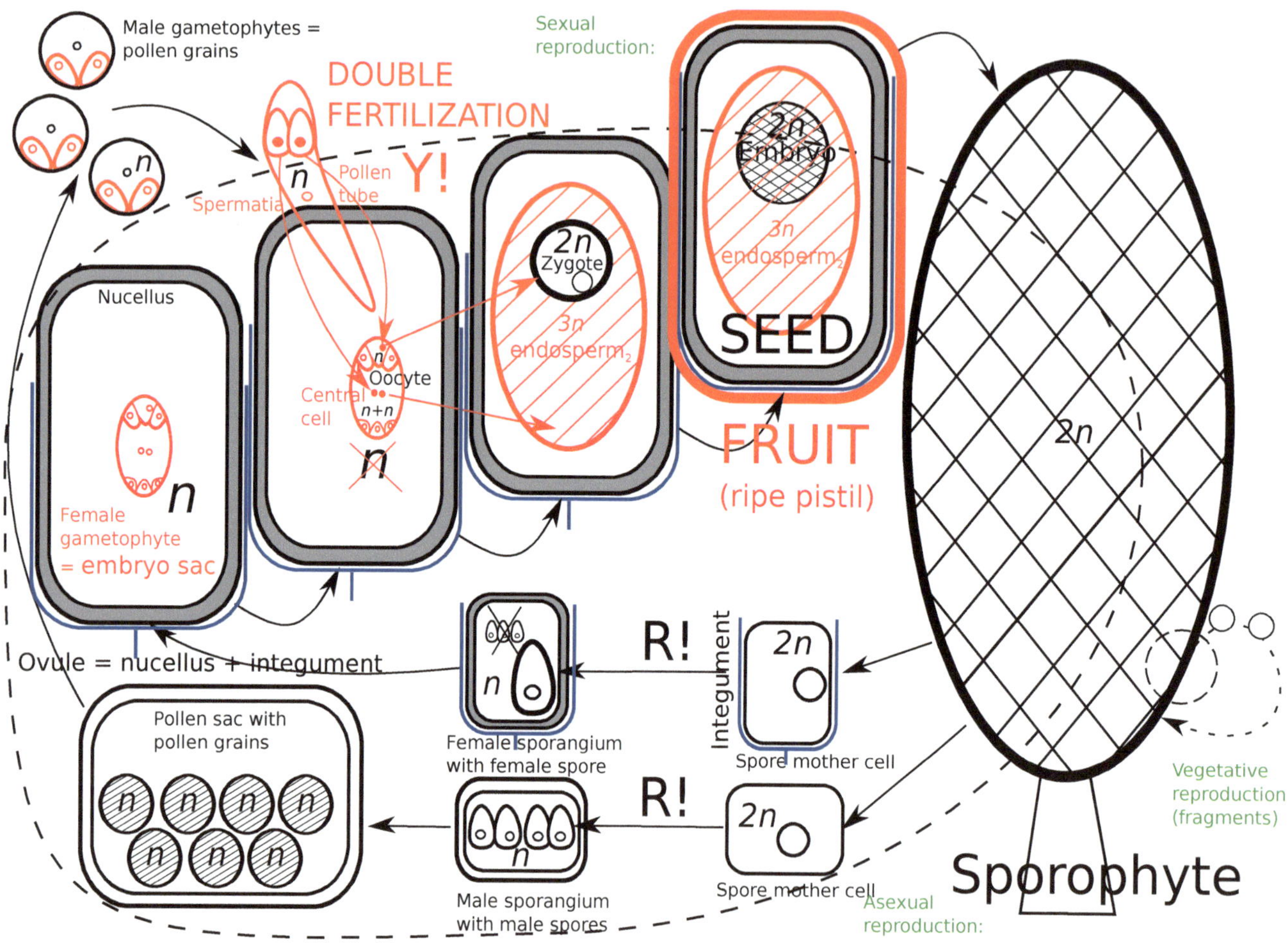

Figure 8.3. Life cycle of angiosperms. Innovations (comparing with the the ancestral seed plant life cycle) are in red.

want to *pollinate* the target plant, and second, they still need to *fertilize* egg cell. Naturally, keeping these two "gunshots" is more complicated then keeping one.

Second shot ancestrally uses water, but higher seed plants managed to get rid of it with pollen tube. First shot used wind which is a natural pollination agent. However, more sophisticated pollination (like insect pollination) was hard to achieve, partly because it requires edible parts like nectar or excess pollen.

If gymnosperms were to increase the speed of life cycle, make more sexual structures, grow rapidly, improve vegetative reproduction, make better pollination and seed dispersal, they could win the competition with ferns in the understory. This is exactly what happened with flowering plant ancestors. Flowering plants grow fast and restore missing (eaten) parts with high speed, they **parcellate** (clone from body parts) easily, they have small and numerous floral units (flowers) which are frequently bisexual but protected from self-pollination and adapted to insect pollination, they guard ovules with pistil wall, their pollen tube grows in hours (not days and weeks), they use fruits to distribute seeds.

Since gymnosperm fertilization occurs *after* gametophyte development, there is frequently a waste of resources: if fertilization does not occur, then all nutrition tissue (endosperm$_1$) will be lost; such empty seeds are unfortunately not rare among gymnosperms. Fertilization of angiosperms involves the *signaling event*: when second sperm fertilizes central cell, it "rings a bell" saying that the first fertilization is now completed. Endosperm (endosperm$_2$ in that case) will start to develop only after the fertilization, and resources will not be wasted. This agile life cycle is the main achievement of angiosperms.

There is a growing evidence that these ancestors were *paleoherbs*, herbaceous plants (and maybe, even water plants like one of the most primitive angiosperms, fossil *Archaefructus*, or basal extant *Ceratophyllum*). Right after they won a competition with herbaceous spore plants, they started to conquer the tree storey again, and now, angiosperms dominate the Earth. There are more than 250,000 species of them which is more than any other group of living beings except insects. There are about 300 families and around 40 different orders. The only places that angiosperms do not grow are the open ocean and the central Antarctic.

* * *

The life cycle of angiosperm (Fig. 8.3) begins much like that of other seed plants; however, when it reaches the point of fertilization, it changes. The male gametophytes, pollen grains, produce pollen tubes which rapidly grow to the ovule and deeper, to the embryo sac. The embryo sac typically has seven cells and eight nuclei (two nuclei in the central cell). The first sperm fertilizes the egg and produces the zygote whereas the second sperm fertilizes the central cell and produces the mother cell of the endosperm$_2$:

1. 1st sperm cell (1st spermatium, n) + egg cell (n) $\rightarrow$ zygote ($2n$)
2. 2nd sperm cell (2nd spermatium, n) + central cell ($2n$ or sometimes n) $\rightarrow$ mother cell of endosperm$_2$ ($3n$ or sometimes $2n$)

(At the time of fertilization, central cell could be haploid, with one nucleus, or diploid, with two nuclei; this is because it runs mitosis without cytokinesis at the end. Consequently, nucleus of the second sperm fuses with either one or two nuclei and endosperm$_2$ is either diploid or (more often) haploid.)

At the end of life cycle, the flowering plant develops the fruit (Fig 8.4). Each part of the fruit is of different origin: fruit skin and wall are from mother plant pistil, seed coat is from mother plan ovule, endosperm$_2$ is a result of second fertilization, and embryo is a daughter plant resulting from the first fertilization.

What is interesting, the embryo of angiosperms is still parasitic: it lives on endosperm which originates from (fertilized, ignited) cell of female gametophyte—in essence, still similar to mosses.

8.2 The Flower and the Fruit

8.2.1 The Flower

A **flower** (Fig. 8.6) is a compact generative shoot that is comprised of three zones: sterile (*perianth*), male (*androecium*), and female (*gynoecium*) (Fig. 8.5). Perianth is typically split into green part (*calyx*, consists of *sepals*) and color part (*corolla*, consists of *petals*). Sometimes perianth consists of similar parts which are neither sepals nor petals: *tepals*. This might be seen in the tulip (*Tulipa*) flower where tepals change their color from green (like in calyx) to red, white or yellow (like in corolla).

The general characters that a flower has are sex, merosity, symmetry, and the position of the gynoecium. **Merosity** is simply the number of parts in each whorl of a plant structure, whether it is the number of sepals, petals in a corolla, or the number of stamens. The position of the gynoecium refers to whether the ovary is superior or inferior (Fig. 8.9). *Inferior ovary* (cucumber, *Cucumis*, apple *Malus* or banana *Musa*) will develop into a fruit where stalk and remnants of perianth are on the opposite ends, whereas *superior ovary* will make fruit where stalk is placed together with perianth (like in tomatoes, *Solanum*). More terms are described in the following separate small glossary:

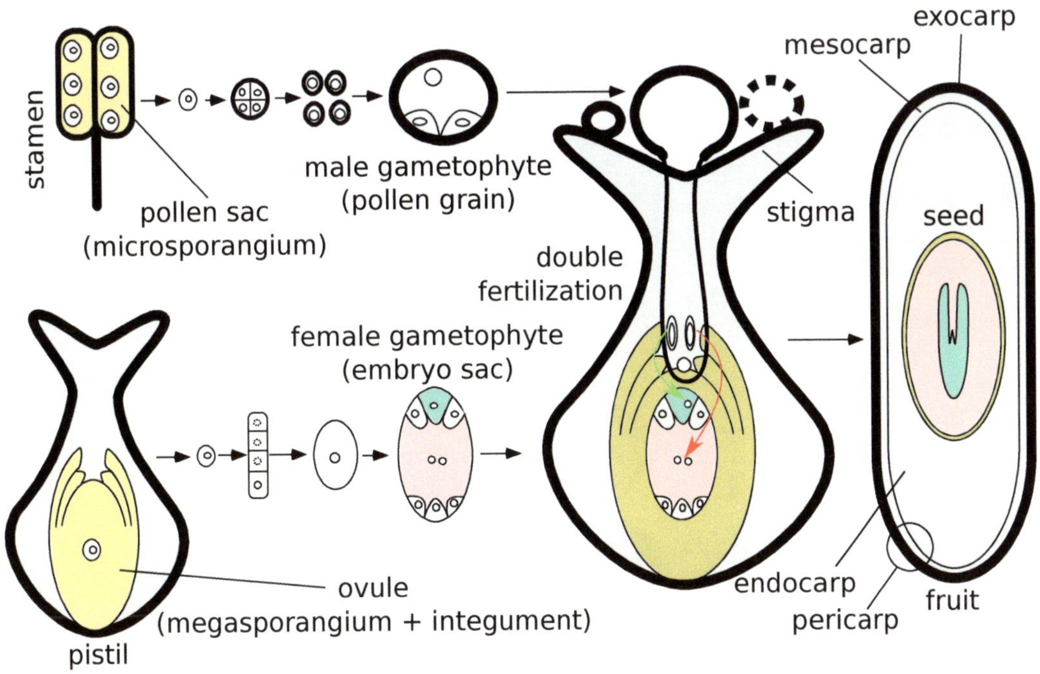

Figure 8.4. The origin of fruit. Note correspondences between different parts (shown with color).

FLOWER PARTS occur in whorls in the following order—sepals, petals, stamens, pistils.

(The only exceptions are flowers of *Eupomatia* with stamens then perianth, *Lacandonia* with pistils then stamens, and some monocots like *Triglochin*, where stamens in several whorls connect with tepals.)

PEDICEL flower stem

RECEPTACLE base of flower where other parts attach

HYPANTHIUM cup-shaped receptacle (Fig. 8.7)

PERIANTH = CALYX + COROLLA

SEPALS small and green, collectively called the CALYX, **formula**: K

PETALS often large and showy, collectively called the COROLLA, **formula**: C

TEPALS used when sepals and petals are not distinguishable, they form SIMPLE PERIANTH, **formula**: P

Figure 8.5. Zones in hellebore (*Helleborus*) flower: sterile perianth, male androecium and in the center, female gynoecium (inside, three ovules are well visible).

ANDROECIUM collective term for stamens: **formula**: A

STAMEN = FILAMENT + ANTHER

ANTHER structure containing pollen grains

FILAMENT structure connecting anther to receptacle

GYNOECIUM collective term for pistils/carpels, **formula**: G. Gynoecium can be composed of:

1. A single CARPEL = simple PISTIL, this is MONOMERY
2. Two or more fused CARPELS = compound PISTIL, this is SYNCARPY
3. Two or more unfused CARPELS = two or more simple PISTILS, this is APOCARPY

(Note that variant #4, several compound pistils, does not exist in nature.)

To determine the number of CARPELS in a compound PISTIL, count LOCULES, points of placentation, number of STYLES, STIGMA and OVARY lobes.

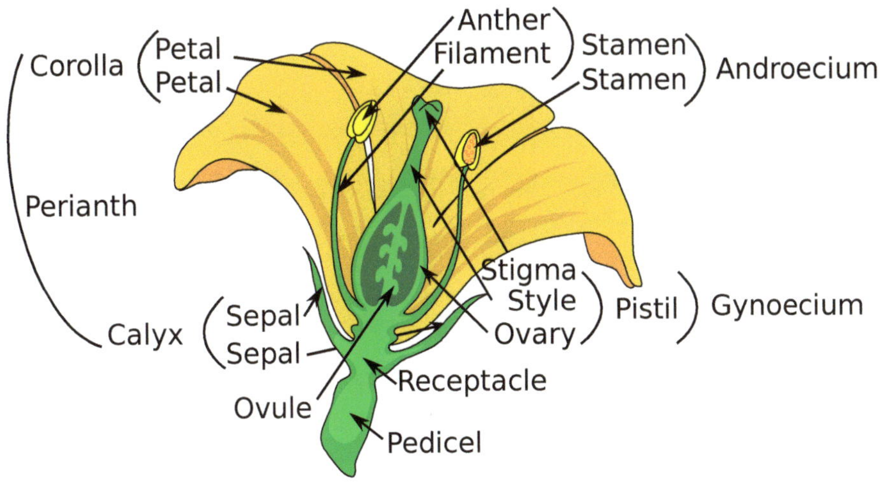

Figure 8.6. Most important parts of the flower.

PISTIL Collective term for carpel(s). The terms CARPEL and PISTIL are equivalent when there is no fusion, if fusion occurs then you have 2 or more CARPELS united into one PISTIL.

CARPEL structure enclosing ovules, may correspond with locules or placentas

OVARY basal position of pistil where OVULES are located. The ovary develops into the fruit; OVULES develop into seeds after fertilization.

LOCULE chamber containg OVULES

PLACENTA place of attachment of OVULE(S) within ovary

STIGMA receptive surface for pollen

STYLE structure connecting ovary and stigma

FLOWER Floral unit with sterile, male and female zones

ACTINOMORPHIC FLOWER A flower having multiple planes of symmetry, **formula**: $\ast$

ZYGOMORPHIC FLOWER A flower having only one plane of symmetry, **formula**: $\uparrow$

PERFECT FLOWER A flower having both sexes

MALE / FEMALE FLOWER A flower having one sex, **formula**: ♂ / ♀ (Fig. 8.8)

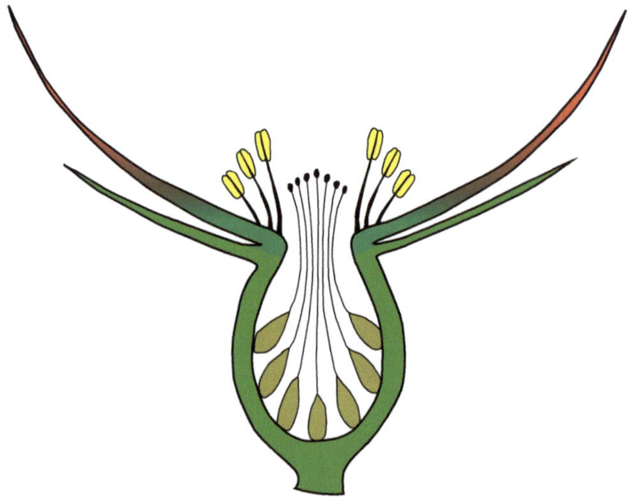

Figure 8.7. Flower with hypanthium (cup-shaped receptacle).

MONOECIOUS PLANTS A plant with unisexual flowers with both sexes on the same plant

DIOECIOUS PLANTS A plant with unisexual flowers with one sex on each plant, in effect, male and female plants

SUPERIOR OVARY most of the flower is attached below the ovary, **formula**: $G_{\underline{\ldots}}$

INFERIOR OVARY most of the flower is attached on the top of ovary, **formula**: $G_{\overline{\ldots}}$

(Inferior ovary only corresponds with monomeric or syncarpous flowers.)

WHORL flower parts attached to one node

Flower formula and diagram

Since there are so many terms about flowers, and at the same time, flower structure and diversity always were of immense importance in botany, two specific ways were developed to make flower description more compact. First is a flower formula. This is an approach where every part of flower is designated with a specific letter, numbers of parts with digits, and some other features (whorls, fusion, position) with other signs:

$\ast K_4C_4A_{2+4}G_{\underline{(2)}}$: flower actinomorphic, with four sepals, four petals and six stamens in two whorls, ovary superior, with two fused carpels

Figure 8.8. Diagram of male (left) and diagram and scheme of female (central and right) flowers of sedges (*Carex*). Note the perigynium (external cover of pistil).

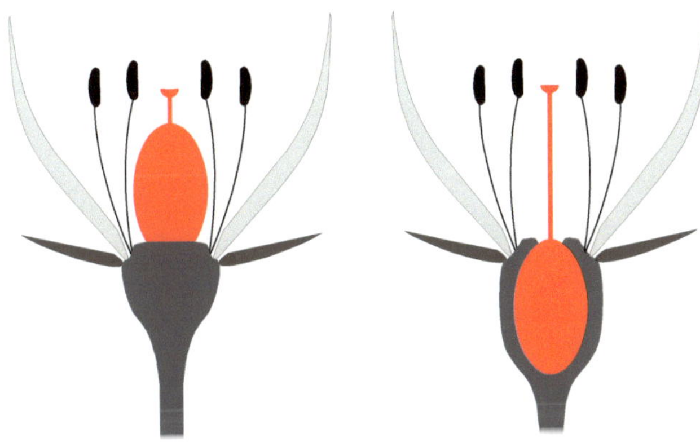

Figure 8.9. Position of ovary: superior (left, *hypogynous flower*) and inferior (right, *epigynous flower*).

> $\uparrow K_{(5)}[C_{(1,2,2)}A_{2,2}]G_{\underline{(2\times 2)}}$: flower zygomorphic, with five fused sepals, five unequal fused petals, two-paired stamens attached to petals, superior ovary with two subdivided carpels
>
> $\ast K_{(5)}C_{(5)}[A_5 G_{\underline{(3)}}]$: actinomorphic flower with five fused sepals and five fused petals, five stamens attached to pistil, ovary inferior, with three fused carpels

The following signs are used to enrich formulas:

PLUS "+" is used to show different whorls; *minus* "–" shows variation; "∨" = "*or*"

BRACKETS "[]" and "()" show fusion

COMMA "," shows inequality of flower parts in one whorl

MULTIPLICATION "×" shows splitting

Figure 8.10. How to draw a diagram (graphical explanation): compare numbers on plant and on diagram.

INFINITY "∞" shows indefinite number of more than 12 parts

Flower diagram is a graphical way of flower description. This diagram is a kind of cross-section of the flower. Frequently, the structure of pistil is not shown on the diagram. Also, diagrams sometimes contain signs for the description of main stem (axis) and flower-related leaf (bract). The best way to show how to draw diagram is also graphical (Fig. 8.10); formula of the flower shown there is $\ast K_5C_5A_5G_{\underline{(5)}}$.

ABC model

All parts of flower have a specific genetic developmental origin explained in the *ABC model* (Fig. 8.11). There are three classes of genes with expression which overlaps as concentric rings, and these genes determine which cells develop into particular organ of the flower. If there are A and C genes expressed, cells will make sepals and pistils. In areas where A and B are active, petals will form; areas where B and C are active are the sites where stamens will appear. A will make a sepal, C will "create" a carpel:

- A alone $\rightarrow$ calyx

- A + B → corolla
- C + B → androecium
- C alone → gynoecium

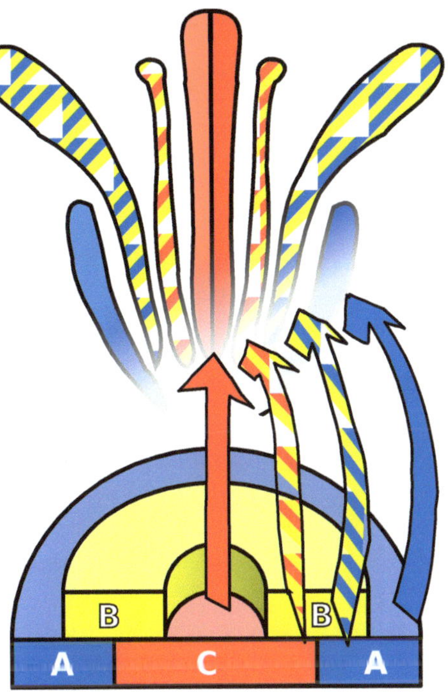

Figure 8.11. ABC model of flower development.

Origin of flower

An example of a primitive magnoliid flower would be *Archaefructus* which is a fossil water plant from the lower Cretaceous time period in China. Its fructifications (flower units, FU) were very primitive and did not yet form a compacted flower, instead, there were multiple free carpels, and paired stamens (Fig. 8.12).

Another ancestral flowering plant is *Amborella*,a small forest shrub of New Caledonia (Fig. 8.13), which is an island in the Pacific Ocean.

Amborella has irregular flowers, a stylar canal, unusual 5-celled embryo sacs that have one central cell, and only four other cells (egg cell and its "sisters"). A stylar canal is a canal that leads to the ovary that the pollen tubes pass through so these plants are not completely "angiospermic", this represents one of the stages of the origin of pistil (Fig. 8.14).

Figure 8.12. Comparison of *Archaefructus* flower (left) and typical flower (note colors). (*Modified from various sources.*)

8.2.2 The Inflorescence

Inflorescence is an isolated generative shoot (shoot bearing FU). Together, inflorescences make **generative shoot system**. Its diverse structure is of not lesser importance than the structure of vegetative shoot system.

The vast diversity of inflorescences can be split into four groups, or "models" (Fig. 8.15). Sole flower is sometimes considered as a "Model 0".

Two models are most widespread. Model I inflorescences are based on the **raceme** (*monopodially branched* generative shoot). They are simple or double and mostly monopodial (Fig. 8.17).

Model II inflorescences (Fig. 8.16) bear or consist of closed (*sympodially branched*) units. The most complete but more rare variant is **thyrsus**, whereas reduced variants (*monochasia* and *dichasia*) are more frequent.

Figure 8.13. *Amborella trichopoda*, sister group to all other flowering plants. White ruler equal to 1 mm.

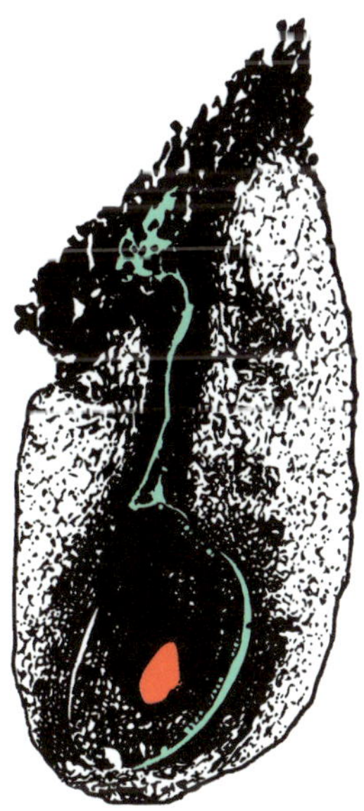

Figure 8.14. *Amborella* pistil, longitudinal section: stylar canal is green, embryo sac red.

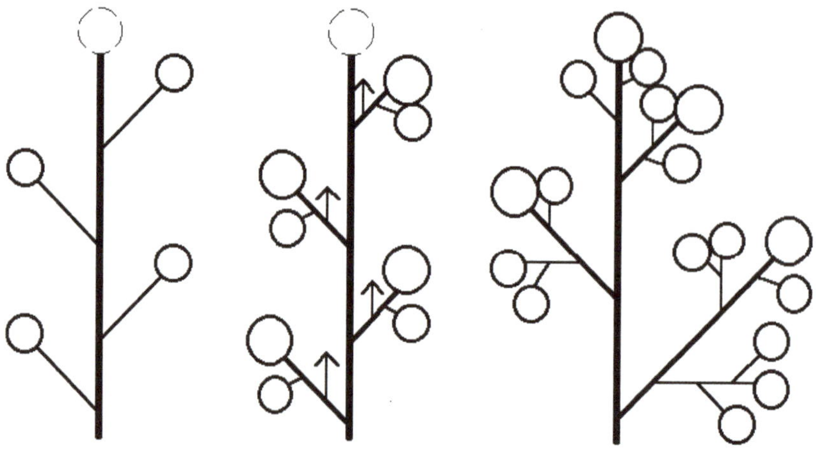

Figure 8.15. Four kinds of inflorescences (left to right): Model I (raceme-based), Model II (thyrsoid) , Model III (panicle) and Model IV (intercalate).

Figure 8.16. Model II inflorescences (from top to bottom): thyrsus, dichasium and monochasium (cincinnus).

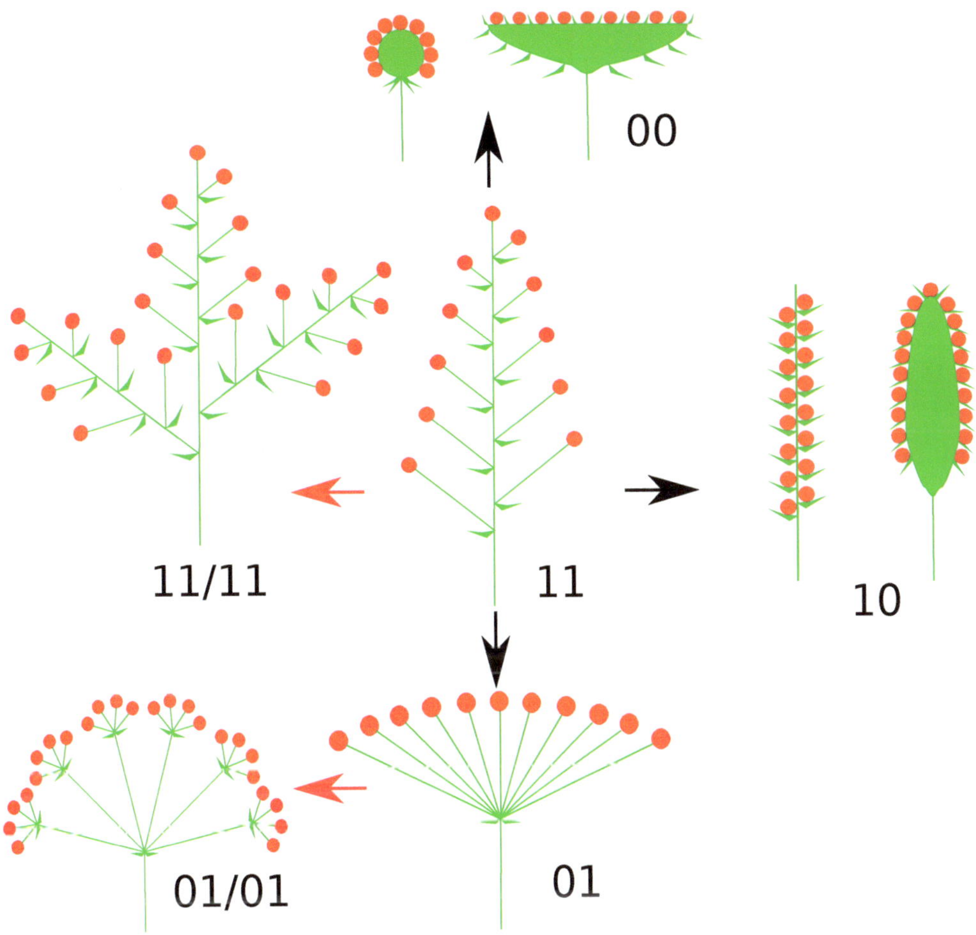

Figure 8.17. Different Model I inflorescences and their evolutionary connections. Digits represent the simple encoding system: first position is main axis, second position are secondary axes (flower pedicels), 1 means developed, 0 reduced. Double inflorescences have four digit positions, for the first and second orders of branching. Some names: 11 raceme, 11/11 double raceme, 10 spike and spadix, 01 umbel, 01/01 compound umbel, 00 head.

8.2.3 Pollination

Pollination could be of two types: self- and cross-pollination. **Cross-pollination** can happen in both abiotic and biotic ways. Abiotic would be represented by gravity, wind, or water; biotic would be performed by agents like insects, birds, bats, or in some cases tree mammals like possums. Wind-pollination is seen as being wasteful and unintelligent due to the fact that the plant needs to produce so much more pollen without any precise targeting.

Adaptation to the particular pollination agent results in different pollination syndromes. For example, cup-shaped flowers are usually pollinated with massive animals like beetles and even bats. Funnel-shaped flowers as well as labiate flowers (with lips), are adapted to flies and bees. Flowers with long spurs attract butterflies and birds (like hummingbirds or sugarbirds).

Self-pollination often exists like a "plan B", in case cross-pollination is, for some reason, impossible. Sometimes, self-pollinated flowers even do not open; these flowers are called **cleistogamous**.

If pollination needs to be avoided, apomixis will prevent it. **Apomixis** requires reproductive organs, but there is no fertilization. One type of apomixis is **apospory** when an embryo develops from the maternal diploid tissue, but does not go through the meiosis stage. In this process, asexual reproduction will have become vegetative. Another type of apomixis would be **apogamy** (parthenogenesis) when embryo develops from an unfertilized gamete after diploidization has occurred. Here, vegetative reproduction evolved from sexual reproduction.

8.2.4 The Fruit

A *fruit* is defined as *ripened ovary, flower, or whole inflorescence*. The origins of the fruit coat and the **pericarp** (Fig. 8.18) which is comprised of the *exocarp*, *mesocarp*, and *endocarp*, are mostly from the wall of the pistil.

Fruits can be simple, multiple, or compound. **Simple fruits** come from a single pistil (like cherry, *Prunus*). **Multiple fruits** are formed from many pistils of the same flower (strawberry, *Fragaria*). A **compound fruit** (infructescense) would be a pineapple (*Ananas*) or fig (*Ficus*) which comes from multiple flowers (inflorescence).

Fruits can be dry or fleshy. An example of *dry fruit* is a nut like peanut (*Arachis*) or walnut (*Juglans*). Examples of *fleshy fruits* include apples (*Malus*) or oranges (*Citrus*).

Fruits also delegate dispersal function to their different parts. **Dehiscent fruits** (like canola, *Brassica*) open and delegate dispersal to individual seeds.

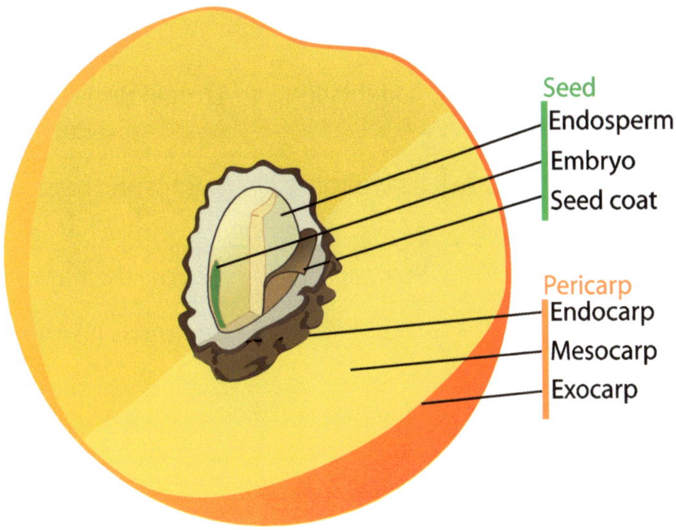

Figure 8.18. Scheme of drupe (e.g., peach) with three levels of pericarp. Note that pit is essentially endocarp + seed.

Indehiscent fruits (like papaya, *Carica*) will not open and will be dispersal units (**diaspores**) themselves.

Schizocarp fruits (like in spurge, *Euphorbia* or maple, *Acer*) are in between: they do not open but break into several parts, and each of them contains one seed inside. For example, maple fruit consists of two "wings", each of them contains the part of fruit and one seed.

In addition, simple fruits could be monomerous (1-seeded) like nut or achene (sunflower, *Helianthus*), or bear multiple seeds (like follicle in tulip, *Tulipa*).

All these different variants have their own names partly described in the following table:

Type	Consistency	Opening	Example(s)
Simple	Fleshy	Indehiscent	Drupe, Berry, Hesperidium, Pome
Simple	Dry	Dehiscent	Capsule, Legume (pod), Silique (Fig. 8.21)
Simple	Dry	Schizocarpic	Regma, Samara, Shizocarp
Simple	Dry	Indehiscent	Caryopsis (grain), Nut (incl. acorn), Achene
Multiple	Fleshy	Indehiscent	Multiple drupe
Multiple	Dry	Dehiscent	Follicle
Multiple	Dry	Indehiscent	Multiple nut
Compound	Fleshy	Indehiscent	Compound berry
Compound	Dry	Indehiscent	Compound nut

8.3 Three plant families you wanted to know but were too afraid to ask

Angiosperms is a giant (quarter of million species) class with four subclasses (Fig. 8.19):

Magnoliidae being the most primitive with flowers of numerous free parts (like water lily, *Nymphaea*, fossil *Archaefructus* and *Amborella*);

Liliidae or **monocots** are grasses, palms, true lilies and many others with trimerous flowers;

Rosidae with pentamerous or tetramerous flowers and free petals;

Asteridae most advanced, bear flowers with fused petals and reduced number of carpels.

Rosids and asterids each comprise about ⅓ of angiosperm diversity.

* * *

Among the numerous taxonomic groups described by scientists in the last 300 years, families of flowering plants hold the distinct place. They were established in col-

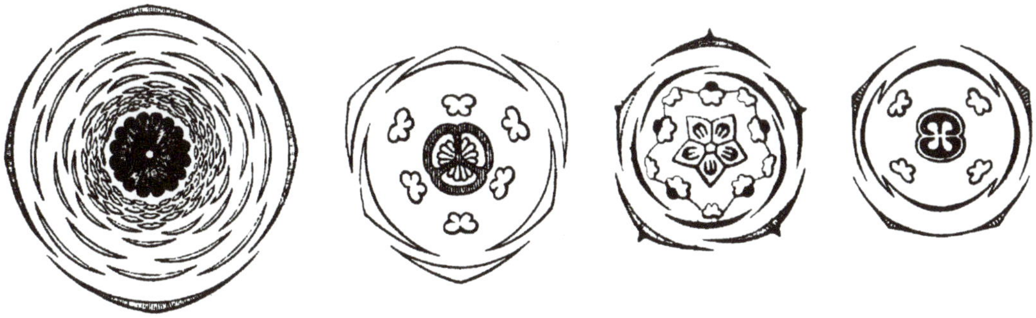

Figure 8.19. Flower diagram "portraits" of flowering plants subclasses (left to right): Magnoliidae, Liliidae, Rosidae, Asteridae. (*Modified from Eichler, 1875*). See also Fig. 8.10 for graphical explanation of the flower diagram.

laborative efforts of French botanists, namely Michel Adanson and Antoine Jussieu. Adanson based his research on methods which are now frequently called "bioinformatics" and therefore was long ahead of his time. Jussieu proved Adanson's ideas by establishing the living garden where plants were arranged by these families. At first, families were not accepted by "fathers of botany" like Carolus Linnaeus. But with time, more and more facts were accumulated which support the ideas enclosed in the families differentiation. The most amazing was almost absolute support of plant families concepts with new molecular methods. Many groups which looked stable (like orders of birds and mammals) appeared less robust than plant families. This is why plant families are so important.

Practically, families provide a great help in knowing plants. For example, the flora of whole North America has 20,000 species of plants. It is almost impossible to remember them all. However, there are only 200 plant families in North America. Therefore, knowing the family saves lots of time and efforts in plant determination.

Several plant families are especially important since they play a big role in economics, form widespread types of vegetation, or are simply extremely rich in species. Three of these families will be characterized below. Characterization of family should follows the plan below:

1. Meta-information: name, position in classification, number of species, distribution
2. Ecological preferences
3. Morphology and anatomy of stem, leaf and root
4. Generative organs from inflorescence to fruit, including flower diagrams and formulas. Seed.

5. Representatives and their importance

8.3.1 Leguminosae, or Fabaceae—legume family

Belong to rosids (Rosidae). Up to 17,000 species, third largest angiosperm family after Compositae (aster family) and Orchidaceae (orchids). Widely distributed throughout the world, but preferably in tropics. Have root nodules with *nitrogen-fixing bacteria*. Leaves alternate, pinnately compound (once or twice), with stipules.

Three subfamilies (Caesalpinioideae, Mimosoideae, Papilionoideae) often treated as separate families. Sepals 5, united. Petals 5, in Papilionoideae they are free, unequal and have special names: *banner*, *keel* and *wing* (Fig. 8.20), in Mimosoideae they fuse and form tube. Stamens often 10 with 9 fused and one free stamen; in Mimosoideae, stamens are numerous. Singe pistil with single carpel. Flower formula of Mimosoideae is

$$\ast K_{(5)}C_{(5)}A_{5-\infty}G_{\underline{1}}$$

Papilionoid legumes have formula like

$$\uparrow K_{(5)}C_{1,2,2}A_{1,[4+5]}G_{\underline{1}}$$

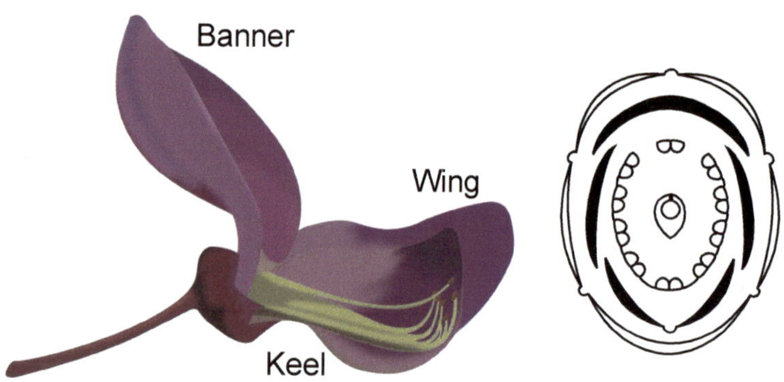

Figure 8.20. Leguminosae. Flowers of Papilionoideae subfamily.

Fruit is a legume (pod): dehiscent with one camera; this is different from silique of cabbage family (Cruciferae) which has two cameras (Fig. 8.21). Mature seeds without endosperm.

* * *

Representatives of Leguminosae:

- Mimosoideae: stamens numerous, petals connected

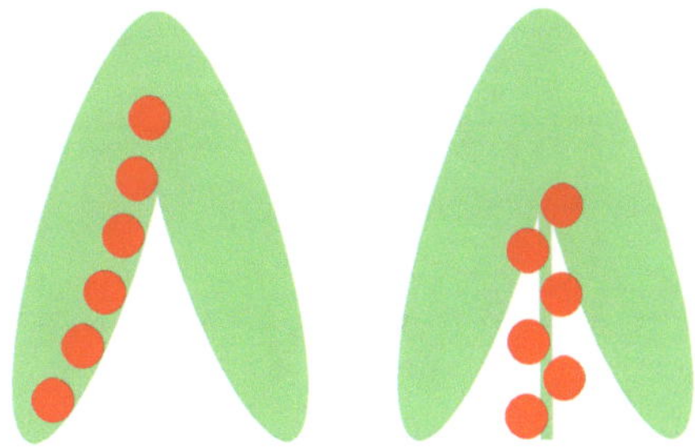

Figure 8.21. Opened pod (Leguminosae) *versus* silique (Cruciferae), seeds are red.

  - *Acacia*—dominant tree of African and Australian savannas, often with phyllodes
  - *Mimosa*—sensitive plant

- Papilionoideae: stamens 9+1, petals free; this subfamily contains many extremely important food plants with high protein value
  - *Glycine*—soybean
  - *Arachis*—peanut with self-buried fruits
  - *Phaseolus*—bean
  - *Pisum*—pea

8.3.2 Compositae, or Asteraceae—aster family

Belong to asterids (Asteridae). More than 20,000 species—second place in flowering plants. Cosmopolitan, but better represented in temperate and subtropical regions. Prefer open spaces. Herbs, rarely woody plants; store carbohydrates as inulin (not starch), sometimes have resin or laticifers (subfamily Cichorioideae). Leaves are alternate or opposite, without stipules, with pterodromous venation.

Flowers in involucrate heads which mimic one flower (Fig. 8.22). Calyx reduced to hairs or bristles (pappus), petals fused in tube or ligula (with 5 or 3 teeth). Stamens 5, fused by anthers, pollen lifted up and distributed by outer sides of stigmas, this is *secondary pollen presentation* (Fig. 8.23). Pistil has 2 carpels, ovary inferior. Fruit is achene, mature seed has almost no endosperm. Flower formula of the tubular (disk) flower is

$$\ast K_{\infty} C_{(5)} A_{(5)} G_{\overline{(2)}}$$

Ligulate (ray) flower typically has formula like

$$\uparrow K_{\infty} C_{(3\vee 5)} A_{(5)} G_{\overline{(2)}}$$

Figure 8.22. Compositae: head and one ligulate (ray) flower.

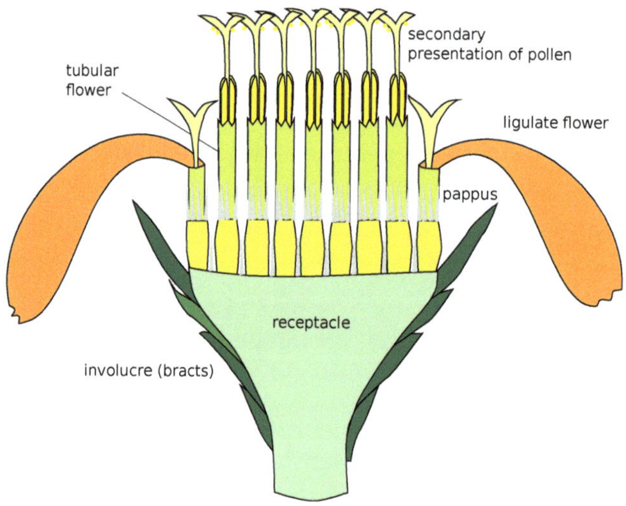

Figure 8.23. Features of Compositae: two types of flowers, secondary pollen presentation, pappus.

Fruit of aster family is one-seeded **achene** (it is a frequent mistake to call it "seed"). In achene, walls of inferior ovary are tightly fused with seed coat. Achenes frequently bear diverse dispersal structures: trichomes, teeth, hooks and others.

* * *

Oil plants, vegetables, ornamentals and medicinal plants distributed in multiple subfamilies, most important are three:

- Carduoideae: mostly tubular flowers
  - *Centaurea*—knapweed
  - *Cynara*—artichoke
  - *Carthamus*–safflower
- Cichorioideae: mostly 5-toothed ligulate (pseudo-ligulate) flowers + lacticifers with latex
  - *Taraxacum*—dandelion
  - *Lactuca*—lettuce
- Asteroideae: tubular + 3-toothed ligulate flowers
  - *Helianthus*—sunflower (BTW, "canola", or Brassica napus from Cruciferae is the second main source of vegetable oil)
  - *Artemisia*—sagebrush
  - *Tagetes*—marigold and lots of other ornamentals

8.3.3 Gramineae, or Poaceae—grass family

Belong to liliids (Liliidae, monocots). Approximately 8,000 species distributed throughout the world, but most genera concentrate in tropics. Prefer dry, sunny places. Often form turf (tussocks)—compact structures where old grass stems, rhizomes, roots, and soil parts are intermixed. Grasses form grasslands—specific ecological communities widely represented on Earth (for example, North American prairies are grasslands). Stems of grasses are usually hollow and round. Leaves with sheaths.

Flowers reduced, wind pollinated, usually bisexual, form complicated spikelets. Each spikelet bears two *glumes*; each flower has *lemma* and *palea* scales (Fig. 8.24). Perianth is reduced to lodicules. Stamens from 6 to 1 (most often 3), with large anthers. Flower formula is

$$\uparrow P_{0-3}A_{0-3+2-3}G_{\underline{(2)}}$$

Fruit is a *caryopsis*; it includes flower scales. Seed contains embryo with *coleoptile*, *coleorhiza* and *scutellum* (Fig. 7.19).

* * *

Most primitive grasses are bamboos (Bambusoideae subfamily). There are many other subfamilies. Two are especially economically important:

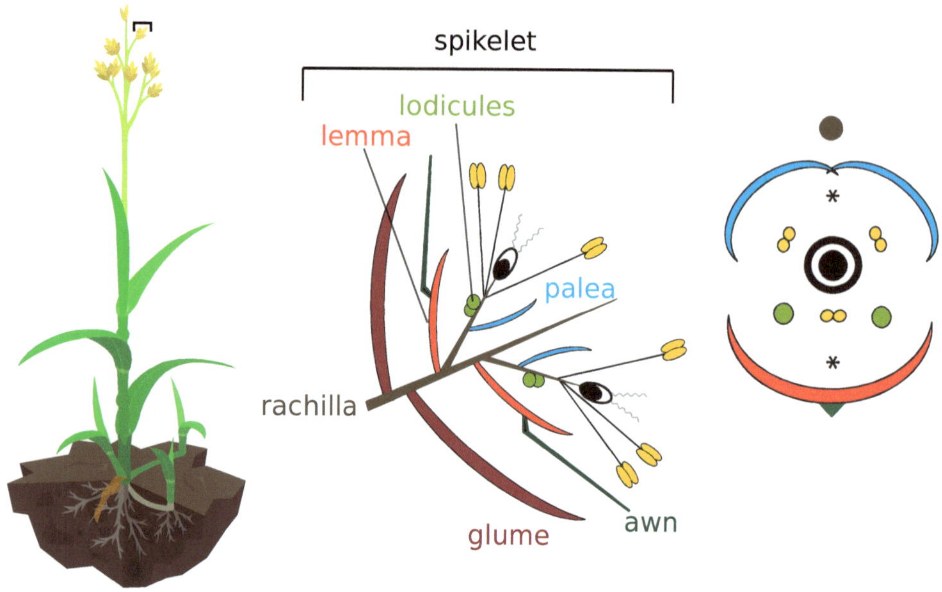

Figure 8.24. Gramineae: one plant, scheme of spikelet and flower diagram.

- Pooid (Pooideae) grasses usually are C_3 plants, wheat (*Triticum*), rice (*Oryza*), barley (*Hordeum*) and rye (*Secale*) belong to this group.
- Panicoid (Panicoideae) grasses are mostly C_4 plants like corn (*Zea*), sorghum (*Sorghum*) and sugarcane (*Saccharum*).

Chapter 9

Plants and Earth

9.1 Geography of Vegetation

Plants are main components of terrestrial ecosystems, they are primary producers, and almost all terrestrial life if based on plants. Consequently, plants will determine how a particular territory might look, which could be, for example, grassland, tundra, or forest. These *types of vegetation* (i.e., visually different plant communities) will have different occurrence on Earth. Below is the list of the most important types (they also called *biomes*):

Tundra Small-sized plants adapted to the short season, wet soils and sometimes also permafrost

Taiga Conifer forests

Deciduous forest Broadleaved temperate forests. The other type of deciduous forests are dry forests of tropical climates.

Grassland Prairie (North America), steppe (Eurasia), savanna (Africa and Australia), llanos (north South America), pampas (south South America)

Shrubland Chaparral (North America), maquis (Mediterranean), fynbos (South Africa), bush (Australia)

Desert Different from shrubland by plants staying apart and soil surface visible

Tropical forest Selva, tropical rain forest: humid and warm environment, the peak of Earth biodiversity

Naturally, these biomes are directly related with the climate, mostly with the coldest temperatures and amount of precipitation. If the Earth would be one continent, then these vegetation types will be arranged from a pole to equator exactly in the order from the list above. However, the real picture is more complicated (Fig. 9.1.)

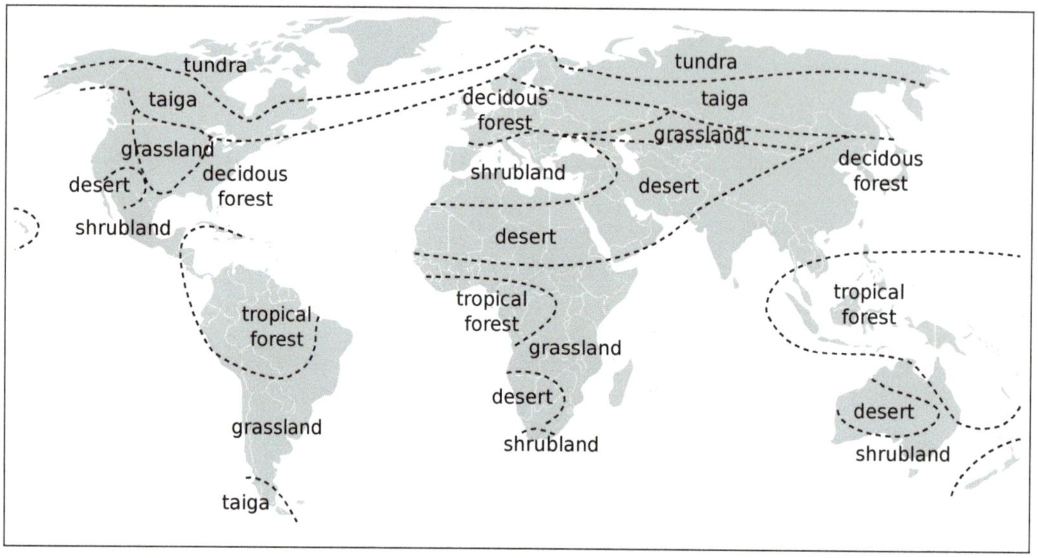

Figure 9.1. Biomes (types of vegetation) of Earth. Please note that this map is largely simplified.

Some smaller biomes, especially different kinds of wetlands (like sphagnum bogs or mangroves) are significantly dispersed, sometimes even *intra-zonal* (occur in different climatic zones).

9.2 Geography of Vegetabilia

While taiga forest looks similar in Alaska (North America) and Patagonia (South America), a closer look will immediately reflect that species, genera and even families of plants are quite different. As an example, both Alaska and Patagonia forests include large conifers, but while in Alaska we frequently see members of Pinaceae family like spruces (*Picea*) or firs (*Abies*), in Patagonia these trees are absent and "replaced" with superficially similar trees of Araucariaceae and Podocarpaceae conifers.

Analogously, Arizona desert is similar to African Kalahari but while American deserts are rich with cacti, similarly looking African plants belong to completely different group, succulent spurges (*Euphorbia*). The effect of these differences on the botanically educated traveler is a bit similar to the nightmare when you first see a familiar thing but approach it—and realize that this is something completely alien and strange.

These *floristic* differences are due to the various geological and biological histories of these places. *Plant biogeography* studies them, explains them and creates the *floristic kingdoms* classification (Fig. 9.2) which takes into account not ecological but taxonomical (phylogenetic) similarities and differences.

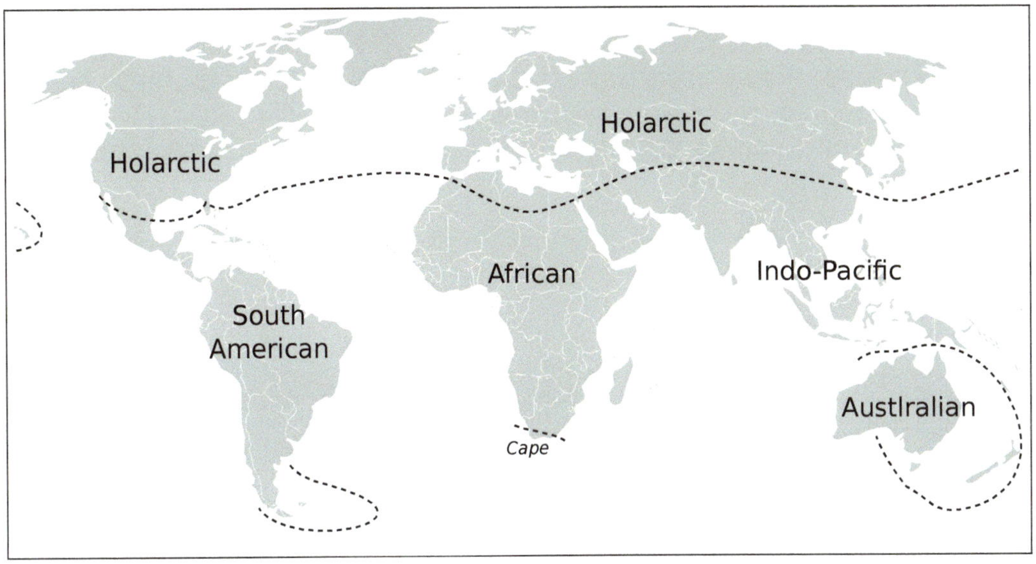

Figure 9.2. Floristic kingdoms of Earth.

There are only five floristic kingdoms:

Holarctic Most of North America and temperate Eurasia. Holarctic kingdom is largest, it covers two continents and most of Northern hemisphere. Typical representatives are pines (*Pinus*) and oaks (*Quercus*).

South American From South Florida to Patagonia and Antarctic islands. Aroids (Araceae family) and bromeliads (Bromeliaceae) are very common South American groups.

African Excluding Mediterranean Africa (very north of the continent). African acacias (*Senegalia*) are common to the most of savannas there.

Sometimes, botanists separate the southern tip of Africa into smallest **Cape floristic kingdom** which has multiple endemic plant genera (like *Berzelia*, kolkol) and even whole families.

Indo-Pacific From India to pacific islands including Hawaii. This kingdom is especially rich of orchids (Orchidaceae); tropical pitcher plants (*Nepenthes*) grow only there.

Australian Australia, Tasmania and New Zealand. Numerous specific plant groups, including *Eucalyptus*, *Banksia* and many others.

Every plant group has a specific *range*—the area of distribution. There are multiple common ranges, e.g., circumpolar (groups distributed across North Pole, both in North America and Eurasia, like spruces, *Picea*) or Gondwanian (groups distributed in the South Africa, Australia and South America, like protea family, Proteaceae). Sometimes, there are *disjunctions* (breaks in range); a typical explanation for the disjunction is long-distance dispersal (like for ispaghula, *Plantago ovata* in California and West Asia) or extinction in the connecting places (like for tulip tree, *Liriodendron* in China and Atlantic states).

Recently, many plants became *invasive* after being introduced willingly (e.g., as forage plants) or accidentally (e.g., with seeds of other plants). These plants (like Eurasian spotted knapweed, *Centaurea stoebe* in North America, or North American box elder, *Acer negundo* in Eurasia) are often *noxious* since they tend to destroy the native vegetation.

It is frequently said that humans stated the new epoch of Earth life, *homocene*—era of *Homo sapiens* dominance, homogenization and great extinction of the flora and fauna. We need to stop that!

Appendix A

Methods of Taxonomy and Diagnostics

The goal of taxonomy is to describe diversity, provide an insight to the evolutionary history (phylogeny), help to determine organisms (diagnostics) and allow for **taxonomic estimations**. The latter means that if we know features of one plant, the taxonomically close one should have similar features. For example, plants from cabbage family (Cruciferae) contain mustard oil (which is responsible for the horseradish taste of many of them). DNA analysis shows that papaya (*Carica* from Moringaceae family) is taxonomically close to Cruciferae. We may guess that papaya also have mustard oil, and this is true! Papaya seeds have the prominent horseradish taste.

* * *

One of the oldest methods of taxonomy is expert-based. Experts produce classifications based of their exclusive knowledge about groups. First taxonomic expert was Carolus Linnaeus (XVIII century). Experts use a variety of methods, including phenetics, cladistics (see below), general evolutionary approach, their ability to reshape available information and their intuition. Their goal is to create the "mind model" of diversity and then convert it to classification, using neighbor groups as a reference (for example, to assign ranks).

A.1 Cladistics

The more contemporary, much more formalized than expert-based is **cladistics**. Below, cladistic procedure is explained using artificial example of three organisms. The goal of the analysis is the creation of a **phylogeny tree** (**cladogram**) which becomes the basis of classification. Below is a short instruction which explains the basics of the cladistic analysis on the artificial example of several "families" of plants.

1. Start with determining the "players"—all subtaxa from bigger group. In our case, it will be these three "families":

   Alphaceae
   Betaceae
   Gammaceae

2. Describe these three groups:

   Alphaceae: Flowers red, petioles short, leaves whole, spines absent
   Betaceae: Flowers red, petioles long, leaves whole, spines absent
   Gammaceae: Flowers green, petioles short, leaves dissected, spines present

3. Determine individual characters (we will need at least $2N + 1$ characters where N is number of studied taxa):

   (1) Flower color
   (2) Petiole size
   (3) Dissection of leaves
   (4) Presence of spines

4. **Polarize the characters**: every character should have at least two **character states** where "0" is ancestral, **plesiomorphic** state, and "1" is derived, **apomorphic** state. To decide which state is plesiomorphic and which is apomorphic, use these kinds of arguments:

   (a) Historical evidence (e.g., from fossils)
   (b) Developmental evidence
   (c) Comparative evidence

5. If this information is absent, find the **outgroup** which is the most ancestral, most early divergent taxon related to our groups. In our case, we will employ outgroup:

   Omegaceae: Flowers green, petioles short, leaves whole, spines absent.

6. Label characters with "1" (apomorphic) or "0" (plesiomorphic):

   (1) Flower color green—0; red—1
   (2) Petiole size small—0; big—1
   (3) Dissection of leaves absent—0; present—1
   (4) Absence of spines—0; spines present—1

7. Make character table containing both subtaxa and labeled characters:

	(1)	(2)	(3)	(4)
Alphaceae	1	0	0	0
Betaceae	1	1	0	0
Gammaceae	0	0	1	1

(Outgroup, Omegaceae evidently has all zeroes.)

8. Start the tree from outgroup (this step is not absolutely necessary but will make phylogeny more clear):

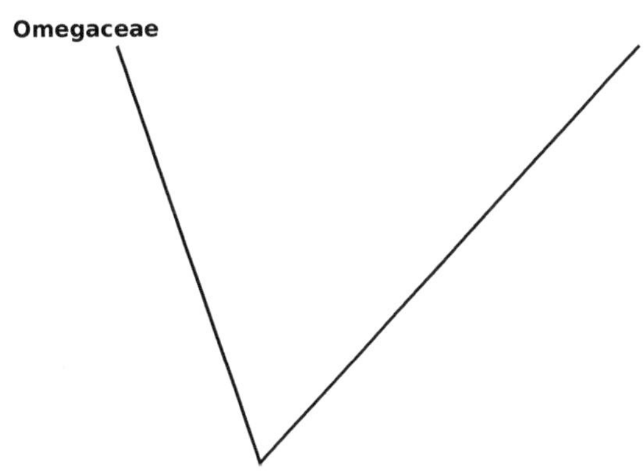

9. Most ancient ingroup (Alphaceae) is a first branch, label it with bar which shows acquisition of the advanced state of first character (red flower color):

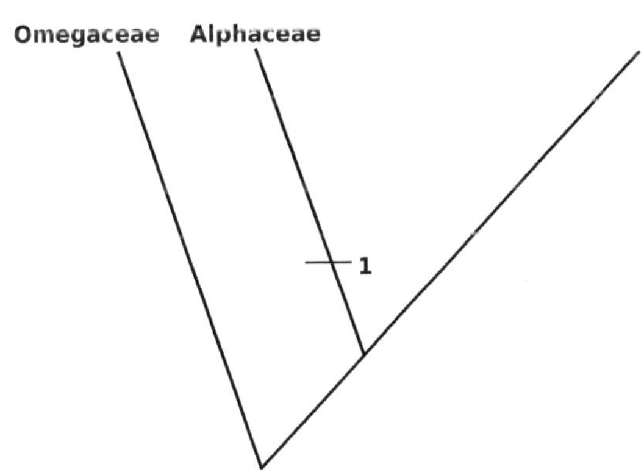

10. Attach more and more sub-taxa. It is possible to do this randomly (like most of phylogeny software), or attach groups to make shortest tree. For example, Betaceae and Gammaceae have equal number of synapomorphies but Betaceae

have only one character different from Alphaceae it is sensible to attach it first, and then attach Gammaceae:

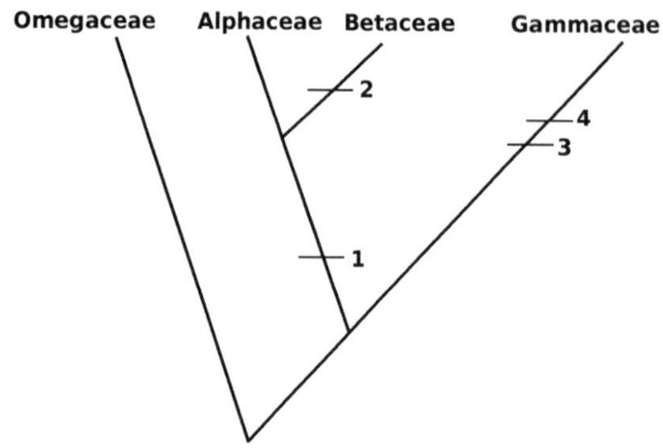

This tree has 4 evolutionary events (length = 4)

11. If Gammaceae was attached first, then resulted tree will be be one step longer:

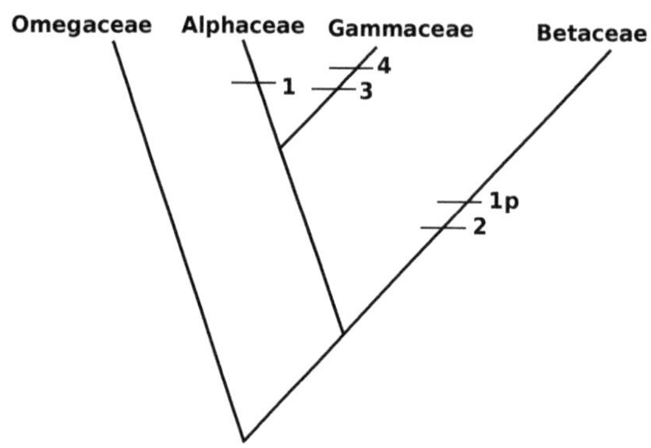

There are five evolutionary events; in other words, length of tree = 5. ("p" are *parallel* characters (homoplasies); there might be also reversals ("r"), when apomorphic character disappears).

There could be also tree with length = 6, or even more if tree includes character reversals, but all of them will be longer than the first one.

12. Choose the shortest, most **parsimonous** tree. Second tree has 5 events, first tree has 4 events, others could be only longer. Consequently, we choose the first tree. By the way, many computer programs do not follow the procedure above strictly and simply produce all possible trees, and finally choose the shortest.

13. Use the chosen tree as a source of classification:

Order Alphales

1. Family Alphaceae
2. Family Betaceae

Order Gammales

1. Family Gammaceae

This step is needed only if you wish to convert cladogram into traditional, classification. In fact, cladograms are rank-free and might be used as is.

Cladograms often used as source of *time trees* which are made with genetic information and information from fossils. If we know the age of taxonomic group, we can use it as more objective replacement of rank.

* * *

Ability to review and compare phylogenetic trees requires understanding of several basic rules, for example:

1. Tree edges may be freely rotated in any direction. For example, these trees are same:

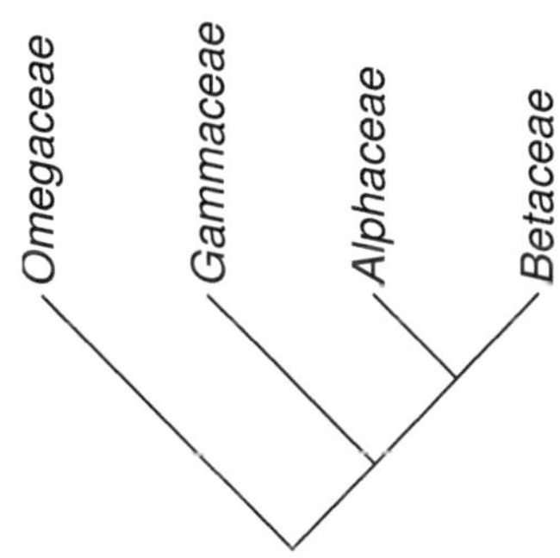

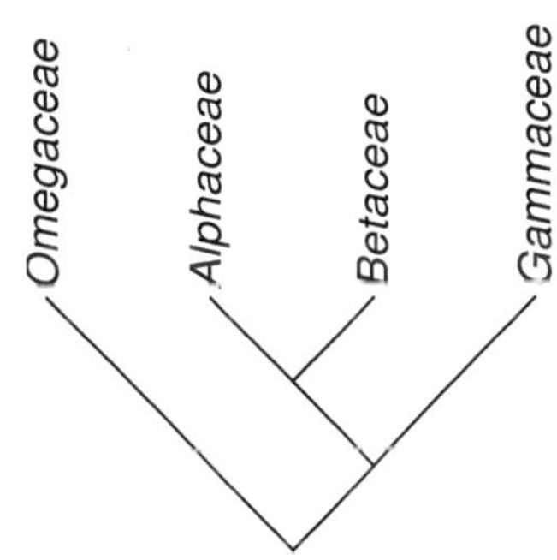

2. Direction of branches also does not matter. These trees are same:

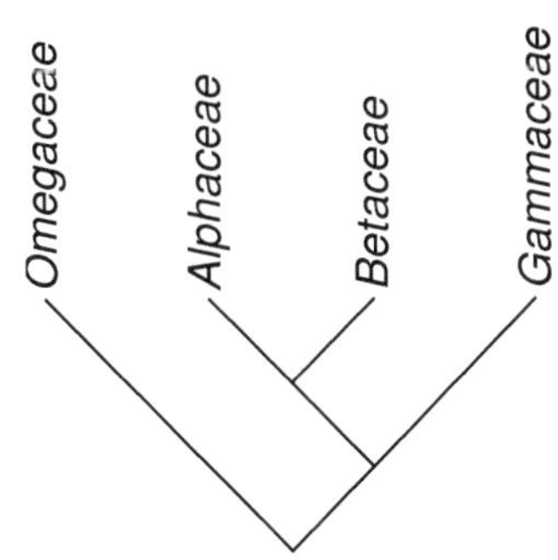

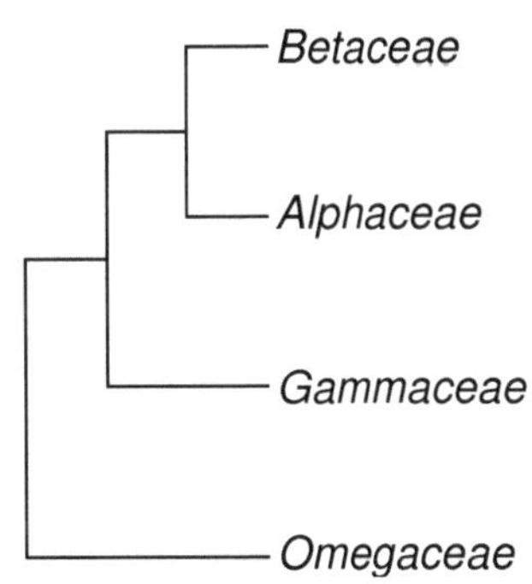

* * *

It is not always simple to make classification from a tree. On the previous example, we simply designate the whole branch as a taxon (order which contains our three families). There are situations when only middle part of the branch seems to be acceptable as a taxon. In these cases, remaining part is called **paraphyletic** taxon.

Paraphyletic groups include all immediate ancestors of its members but not all descendants of these ancestors. Good example of paraphyletic taxon are reptiles: when we take mammals and birds from amniote branch, reptiles will be what is left. Gymnosperms (all seed plants without angiosperms) is another example, but in this case *some* molecular trees show that gymnosperms is also a natural branch (i.e., **monophyletic** group). Monophyletic groups include all immediate ancestors and all their descendants.

When the group contains taxa from different branches, it is **polyphyletic**. Polyphyletic groups are not allowed.

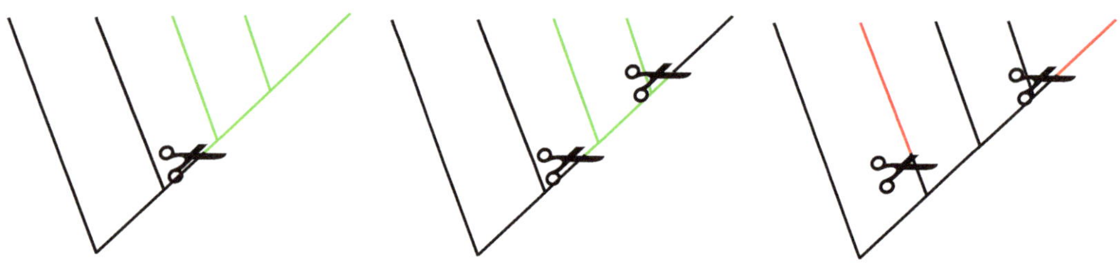

Figure A.1. Monophyletic, paraphyletic and polyphyletic groups, respectively.

Another important distinction between groups of the phylogeny tree is stem and crown groups. All extant members together with their immediate ancestors form a **crown** of taxon (Fig. A.2). If one member of crown went extinct, we can estimate that it was somehow similar to other crown members. In other words, if we find a way ho to re-create mammoth, we probably understand how to feed it because it belongs to the Elephantidae family crown. However, if the fossil, extinct members of taxon branch outside of crown (**stem** groups), there are much less taxonomic estimations. It is hard to guess, for example, how to care for *Archaeopteryx* "dinosaur bird" because such organisms are not exist now and have no living similarities.

Figure A.2. Crown (green skull) and stem (red skull) extinct groups among extant groups.

A.2 Phenetics

The other way of making classification is even more mathematical. This is **phenetics** based on multivariate methods of data analysis. One of its methods is **cluster analysis** which is described below.

1. Contrary to cladistics, phenetics considers characters as **all equal** and **does not employ any evolutionary assumptions.**
2. We need to decide which taxa we will need, assess their descriptions, extract characters—all these is similar to cladistics (see above).
3. Character polarization is not needed, character codes may be specified more or less arbitrarily, and there is no need for outgroup.
4. Character table could be the same as in previous example (again, see above).
5. Then, we will need to create the square matrix (or table) of similarity:

	Alphaceae	Betaceae	Gammaceae
Alphaceae	1		
Betaceae	0.75	1	
Gammaceae	0.25	0	1

Every cell of this matrix contains a value of similarity K:

$$K = \frac{\text{number of matching characters}}{\text{number of all characters}}$$

Please note that there are many more relevant coefficients of similarity but they are out of our scope.

6. Then we need to make the dendrogram which is a tree-like structure. Traditionally, dendrogram is built from bottom to top, from more similar to less similar groups. For example, we may start from connecting the closest taxa, Alphaceae and Betaceae:

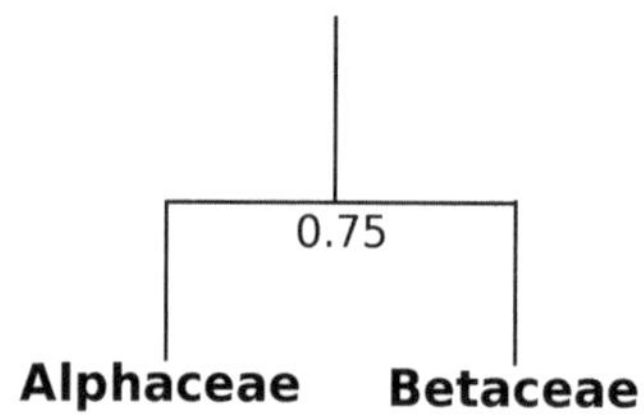

7. Then we need to attach other taxa which are closest to previous group:

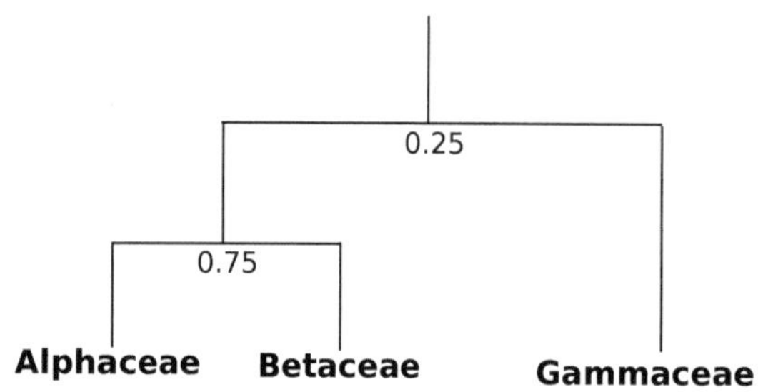

   Sometimes, when we have multiple taxa, we end up with several independent groups (clusters). In that case, different clusters could be connected on the base of *average similarity*.

8. Betaceae and Alphaceae are closer, so we can unite them in one order:

   Order Alphales

   1. Family Alphaceae
   2. Family Betaceae

   Order Gammales

   1. Family Gammaceae

A.3 Dichotomous keys

Diagnostics is a practical science which helps to determine living organisms. One of the best way of determining was invented in the end of 18 century by famous French naturalist, Jean-Baptiste Lamarck. He created the **dichotomous key** (sometimes called descriptive key, or descriptive table). The legend says that when Lamarck demonstrated this key for the first time, he gave it to the random stranger (who had no idea about plants and their names), and plant were determined without problems! How to make such a key? The example is below:

1. We need to start with "players". In this example, it will be same three plant families:

   Alphaceae
   Betaceae
   Gammaceae

2. Assess descriptions of these three groups (we copy this from the above):

   Alphaceae: Flowers red, petioles short, leaves whole, spines absent
   Betaceae: Flowers red, petioles long, leaves whole, spines absent
   Gammaceae: Flowers green, petioles short, leaves dissected, spines present

3. Start with a character which let to split the list into two nearly equal groups. Then add other character(s). It is always good to use more characters!

   1. Petioles long .... **Betaceae.**
   – Petioles short .... 2.

   2. Flowers red, leaves whole, spines absent .... **Alphaceae.**
   – Flowers green, leaves dissected, spines present .... **Gammaceae.**

As you see here, key consists of steps. Every step has a number and typically two choices. Number is attached to the first choice whereas the second choice is marked with minus "–". The choice will lead either to the name, or to another step. The choice sentence might contain several phrases, the first is the most important and the last is the least important.

Appendix B

Problems

Below is the set of botanical problems, questions which require careful thinking and analysis of multiple hypotheses. Please remember that it is rare in biology to have just one answer, so most of questions below have many answers.

1. Some asters and other flowering plants grow in the tidal zone of the sea: during the high tide, they are fully covered with seawater; and during the low tide, they are fully emerged. Please guess morphological and physiological features which allow them to grow in these conditions.

2. Most of the flowers do not have odor or have odor pleasant to us. At the same time, insects like smells of corpses, feces, etc. However, only a few flowers use these last odors. Why the majority of flowers do not use odors unpleasant to us?

3. In 1894, Herbert Walles published the short story entitled "Flowering of the Strange Orchid". In this story, the gardener started to cultivate the mysterious orchid, which was brought from some tropical jungles. One day, he did not return from his greenhouse, and his housekeeper went to find him:

    > "He was lying, face upward, at the foot of the strange orchid. The tentacle-like aerial rootlets no longer swayed freely in the air, but were crowded together, a tangle of grey ropes, and stretched tight, with their ends closely applied to his chin and neck and hands.
    >
    > She did not understand. Then she saw from one of the exultant tentacles upon his cheek there trickled a little thread of blood.
    >
    > With an inarticulate cry she ran towards him, and tried to pull him away from the leech-like suckers. She snapped two of these tentacles, and their sap dripped red.

> Then the overpowering scent of the blossom began to make her head reel. How they clung to him! She tore at the tough ropes, and he and the white inflorescence swam about her. She felt she was fainting, knew she must not."

Which else adaptations (except already described above) will be necessary for orchids to start feeding on the blood of mammals? Maybe, some of the above features should be changed?

4. Wind is hazardous for epiphytic plants that grow high on the trees. The fallen plant might survive below but surely will not be able to flower. How do epiphytic plants withstand the wind?

5. Sometimes, scientists find fossil flowers. How to know how these flowers were pollinated?

6. In lower stories of the tropical forest, there are many plants with striped or spotted leaves—with white, red, yellow spots, or stripes. Especially frequently, these leaves belong to one of the plant families, Marantaceae (by the way, they are frequently cultivated indoors). What (do you think) is the reason for the plant to have such leaves?

7. There are no seasons in humid tropical forests, but trees drop their leaves even there. Some trees drop leaves of the last year(s) together whereas other trees drop leaves one by one, every day or week. What are the benefits of each of these strategies?

8. Water plant quillwort is the direct descendant of giant Paleozoic tree-like lycopods. Like its ancestors, it has an ability to secondary thicken its stem. It is easy to understand why secondary thickening is good for trees. However, is it beneficial for the small water plant, or is this just a "vestigial" feature? Please justify your answer.

9. On the Figure B.1, there are two real plants and three "chimeras" made out of fragments of other plants. Which plants are real? Why? Please explain.

10. In many popular books, it is said that reproduction with seeds is in all aspects better than reproduction with spores. However, in the World flora, the number of species of the spore plants is only 5–7 times less than the number of species for seed plants. Which adaptations of spore plants allow them to keep the significant positions in the World flora?

11. Floras of Eurasia and North America often unite in one Holarctic floristic kingdom, because they have many common genera and even species of plants. Interestingly, some genera (e.g., horsetails, blueberries, nettles, anemones, bluegrasses, clovers, wintergreens, bellflowers) have many common species

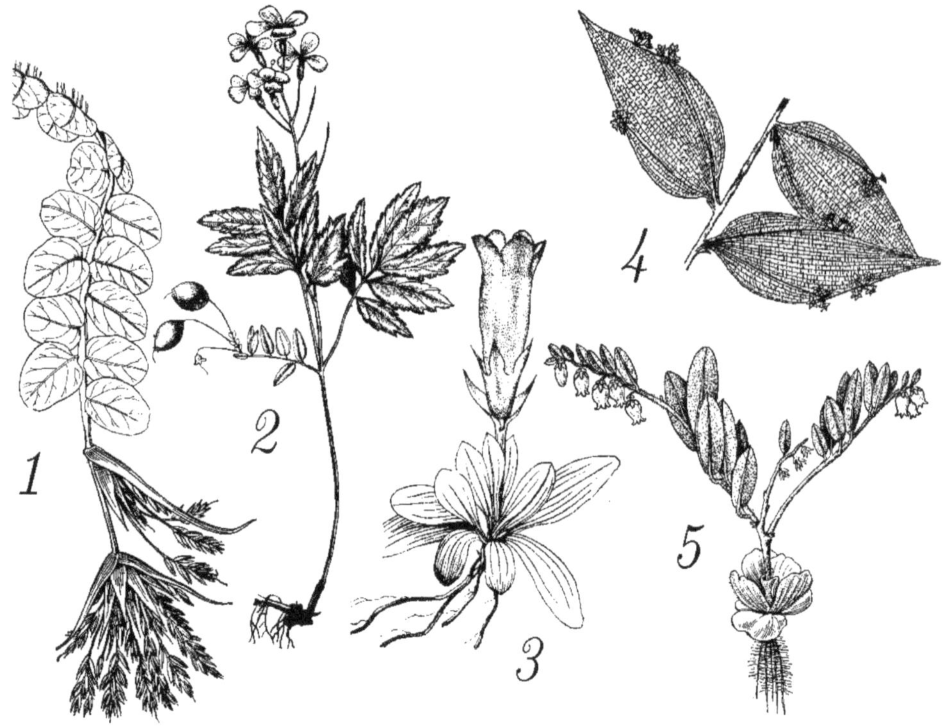

Figure B.1. Real plants and plant chimaeras

on both continents. In contrast, species from other genera (asters, needle grasses, wormwoods, sedges, roses, willows) are different on each continent. Furthermore, there are genera that grow only in either North America or Eurasia (beard-tongues, opuntias, knapweeds). What do you think, what are the reasons of each type of distribution?

12. Some representatives of two families of insect coccids have embryos developed from two different kinds of cells. First kind is just a normal zygote, but the second is the result of the fusion between polar body (meiosis "remain") and one of the daughter nuclei of a zygote. As a result, the second cell is polyploid, frequently with five or even seven sets of chromosomes (this is because the second cell might additionally duplicate its genome). Most organs of the adult insect formed by the first cell, the second cell, gives "bacteriome", the kind of fat tissue which cells bear symbiotic bacteria. What do you think, why coccids have such a complicated process? What are benefits from the presence of two different genotypes in one body?

13. On some islands, there are woody plants that belong to groups that are generally herbaceous (e.g., woody plantains on Hawaii, woody bellflowers on St. Helena). What do you think are reasons for this phenomenon?

14. Within populations of North American painted wintergreen (*Pyrola picta*), there are plants with normal green shoots and also plants with shoots without chlorophyll. These last shoots feed on soil fungi ("mycoheterotrophy"). What are the benefits of this populations structure for the species? Why, in each generation, there are two types of shoots?

15. Some flowers in nature are "double", with increased (comparing with norm) number of petals. How are double flowers beneficial to plants? Which problems might be related to having double flowers?

16. Everybody knows the sensitive plant (*Mimosa pudica*); this is the tropical weed, which after touching its leaves, folds segments and then also moves whole leaves down. Mimosas, however, is a big genus, and some Mimosa species do not react to the touching, some only fold segments (and did not lower leaves), and some do exactly as *M. pudica*, sometimes slower, sometimes faster. What do you think are the advantages and disadvantages of three different sensitivity types above? And what particulars of the plant life promote the sensitivity? Please develop several versions.

17. Some plants are evergreen, and some are deciduous. There are several different strategies of being deciduous: some lose all their leaves literally for one night, some keep falling leaves for few months, and some do not drop leaves in a fall but keep them all winter and drop leaves in spring. What are the advantages of each strategy? Please justify your answer.

18. Every fall, we see fallen leaves. Which other parts plants can drop before the disadvantageous season? Please give some examples.

19. Everybody knows that sunlight is useful for plants. However, gardeners also know that light could be harmful. How exactly could the light of Sun be harmful for plants?

20. Under cover of tropical forest, flowers are not frequent, and frequently plants of the same species flower at different times of the year. Therefore, the distance between simultaneously flowering plants of the same species could be really big. What are the advantages and disadvantages of this flowering rhythm? Please justify your answer.

21. On the Figure B.2, there are different types of plant trichomes (please note that "E" is the top view). What role play each of these trichomes in plant life? Why? Please explain.

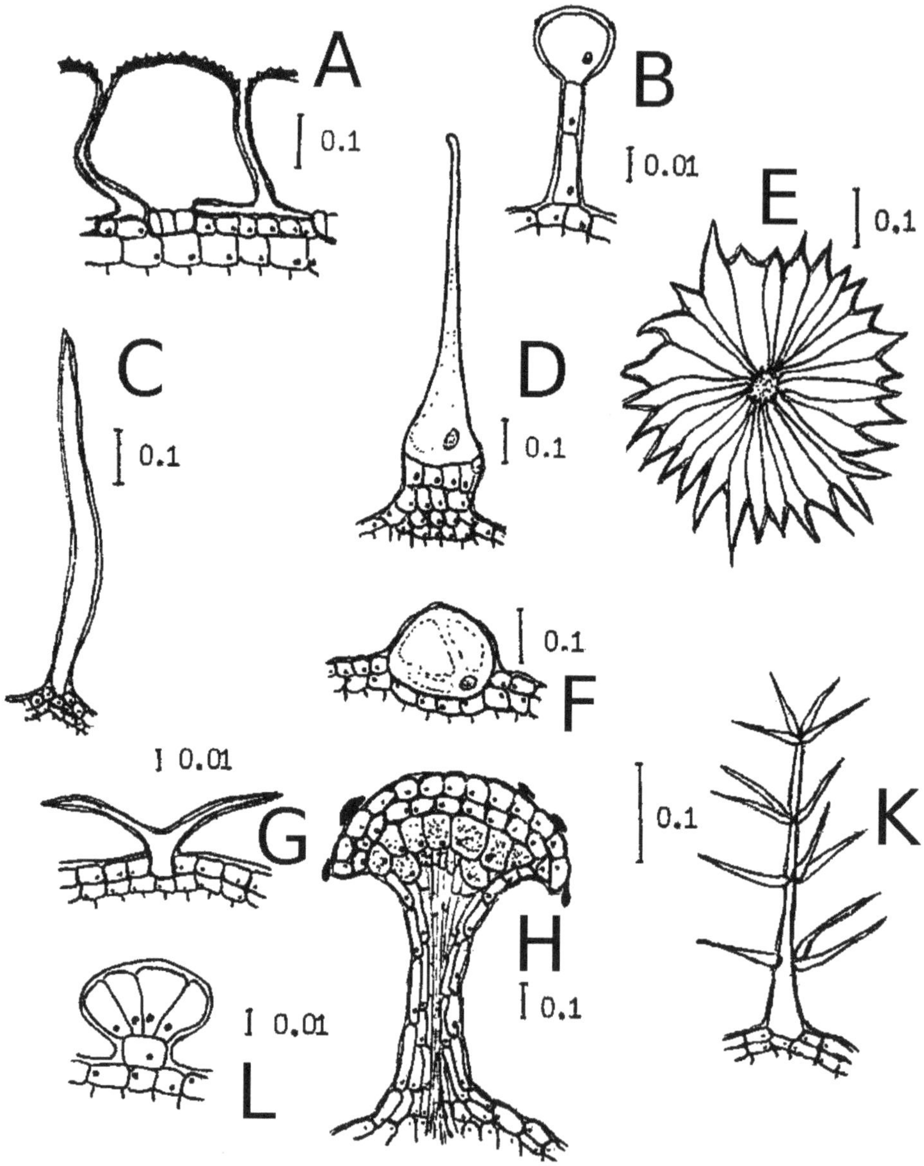

Figure B.2. Plant trichomes

Appendix C

Some useful literature

There are botanical publications which I especially like, and also think that they might be useful to the readers of this book. There are so many of them! But I tried to shorten this list as much as possible. This list is below.

Boudouresque C.F. 2015. Taxonomy and phylogeny of unicellular eukaryotes. In Environmental Microbiology: Fundamentals and Applications (pp. 191–257). Springer, Dordrecht.

Bresinsky A., Körner C., Kadereit J.W., Neuhaus G., Sonnewald U. 2013. Strasburger's plant sciences: including prokaryotes and fungi (Vol. 1). Berlin, Germany: Springer.

Chamovitz D. 2012. What a plant knows: a field guide to the senses. Scientific American/Farrar, Straus and Giroux.

Crang R., Lyons-Sobaski S., Wise R. 2018. Plant Anatomy: A Concept-Based Approach to the Structure of Seed Plants. Springer.

Eichhorn S.E., Evert R.F., Raven P.H. 2012. Biology of plants. WH Freeman & Company.

Gago J., Carriquí M., Nadal M., Clemente-Moreno M.J., Coopman R.E., Fernie A.R., Flexas J. 2019. Photosynthesis optimized across land plant phylogeny. Trends in Plant Science.

Gray A. 1878. Botany for young. Ivison, Blakeman and Company.

Holttum R.E. 1954. Plant life in Malaya. Longmans.

Jäger E., Neumann S., Ohmann E. 2015. Botanik. Springer-Verlag.

Kraehmer H., Baur P. 2013. Weed anatomy. John Wiley & Sons.

Manetas Y. 2012. Alice in the land of plants: biology of plants and their importance for planet earth. Springer Science & Business Media.

Olson M.E., Rosell J.A., Zamora Muñoz S., Castorena M. 2018. Carbon limitation, stem growth rate and the biomechanical cause of Corner's rules. Annals of Botany. 122: 583–592.

Prusinkiewicz P., Lindenmayer A. 2012. The algorithmic beauty of plants. Springer Science & Business Media.

Sage R.F., Monson R.K., Ehleringer J.R., Adachi S., Pearcy R.W. 2018. Some like it hot: The physiological ecology of C_4 plant evolution. Oecologia. 187: 941–966.

Trewavas A. 2014. Plant behaviour and intelligence. OUP Oxford.

von Denffer D., Bell P.R., Coombe D. 1976. Strasburger's textbook of botany. Longman.

Watts W.M. 1910. The school flora for the use of elementary botanical classes. Longmans.

Xu H.H., Berry C.M., Stein W.E., Wang Y., Tang P., Fu Q. 2017. Unique growth strategy in the Earth's first trees revealed in silicified fossil trunks from China. Proceedings of the National Academy of Sciences. 114: 12009–12014.

www.ingramcontent.com/pod-product-compliance
Lightning Source LLC
LaVergne TN
LVHW071242210926
848486LV00007B/326